CONTENTS 目录

What's New

第一空间 VOL.*2

主编
宋纯智

编辑
陈慈良
谢昕宜

流程编辑
刘沉地

设计总监
迟 海
赵 聪

翻译
王晨晖
李 婵

编委
梁建国 赵寂蕙 陈 涛
刘涤宇 李祖原 王茂乐
Alexander Plajer
Heizog & de Meuron
Kelly Hoppen
Luca Scacchetti
Plajer Franz
Randy Brown

市场拓展
李春燕 lchy@mail.lnpgc.com.cn

广 告
杜丙旭 dubx@lnpub.com

印刷/制作
广州恒美印务公司

发行
李大鹏 benli0821@163.com (86 24)2328-0366

读者服务
何桂芬 fxyg@mail.lnpgc.com.cn (86 24)2328-4502

图书在版编目（CIP）数据

第一空间. 2 /《第一空间》编辑组编.—沈阳：辽宁
科学技术出版社，2010.3
ISBN 978-7-5381-6324-7

I. ①第… II ①第… III. ①办公室—室内设计
IV. ①TU24

中国版本图书馆CIP数据核字（2010）第030929号

出版发行：辽宁科学技术出版社
　　　　　（地址：沈阳市和平区十一纬路29号 邮编：110003）
印 刷 者：恒美印务（广州）有限公司
经 销 者：各地新华书店
幅面尺寸：225mm×300mm
印 张：22
插 页：4
字 数：100千字
出版时间：2010年3月第1版
印刷时间：2010年3月第1次印刷
责任编辑：陈慈良
封面设计：迟 海
版式设计：迟 海
责任校对：周 文

书 号：ISBN 978-7-5381-6324-7
定 价：128.00元

联系电话：024-23284360
邮购热线：024-23284502
E-mail:lnkjc@126.com
Http://www.lnkj.com.cn
本书网址：www.lnkj.cn/uri.sh/6324

Architects ONG&ONG have completed a contemporary open-plan interior for the renovation of a terrace house in Singapore, merging the interior and exterior spaces.

55 Blair road project is a renovation and restoration to a traditional Art Deco style shop house, to create a light open-plan living space, whilst promoting inside/outside space. The contrasting relationship between the metallic elements and subtle tones within the house helps create an exciting spatial relationship throughout.

ONG&ONG设计公司完成了新加坡一幢别墅的重新装修，富有现代气息的开敞空间将内外空间联系在一起。

Zurich and London designers Aekae have completed the interior of a café in a Zurich park.

The designer says: "The location at the park offers nice views of the trees, and we didn't want to distract from these with a flashy, overdesigned interior, thus keeping it simple. Inspiration came from classic elements of a French park-café that was interpreted in a modern way. The main objective of the design was to create an unpretentious and relaxed space that effortlessly combines the old and the new." (http://www.aekae.com/)

苏黎世和伦敦设计师Aekae完成了一个苏黎世公园中的咖啡馆室内设计。

Architect Corvin Cristian has designed an office interior inside the former Romanian Stock Exchange Building in Bucharest, Romania.

In the old Romanian Stock Exchange Building, the shipping-case-themed furniture acts as storage, dividing walls, dynamic company statement and reverence on the genius loci. The meeting room has the proportions of a shipping container. The chesterfields and the oversized lamps bring a homey feeling to the otherwise austere design.

Corvin Cristian建筑事务所完成了一个办公室内设计。

Berlin designer Werner Aisslinger has completed a budget hotel designed to feel like staying at a friend's house.

They have created a cosmopolitan, yet street-savvy hangout, where they themselves would love to stay. With rates starting at 59 EUR per night, dreamers, movers and shakers, lovebirds, soul searchers, artists, craftspeople, globetrotters and Berlin aficionados alike can enjoy the special social dynamic of the 250-bed space, which will soon become a Berlin landmark for the young and the young at heart.

柏林设计师Werner Aisslinger完成了一个酒店的设计。

Beijing architects MAD have completed the first of a series of proposed bubble-shaped additions to traditional hutongs in the city.

The hutong bubbles, inserted into the urban fabric, function like magnets, attracting new people, activities, and resources to reactivate the entire neighborhoods. They exist in symbiosis with the old housing. In time, these interventions will become part of Beijing's long history, newly formed membranes within the city's urban tissue.

北京MAD建筑公司开始了为传统北京胡同增加气泡形状的系列工程，如今已经完成了第一个。

Singapore designers Ministry of Design have completed an office interior for an advertising agency.

In the Space to Impress, visitors and guests exiting the lift into the entry foyer are immediately confronted with a larger than life "graffiti" style portrait of Leo, an over 3-metre-high mural painted on the floor, walls, windows and ceiling of the main entry foyer. (http://www.modonline.com)

新加坡设计部为一个广告经销处做了一个办公室内设计。

Shanghai-based design studio EXH Design has completed the extension to the visa savilion of the Swiss Embassy in Beijing.

The designer was approached to enlarge the working areas and provide a fresh representation of Switzerland in China. The concept is to "see the history through the decade". The designers kept the old brick building, which witnessed the accelerated pace of change in China; and extended it with a lightweight glass façades, through which the inside bricks still can be seen. Two kinds of architecture are overlapped across time.

上海EXH设计事务所完成了位于北京的瑞士大使馆签证中心的扩建工程。

German architects Spine have converted a brewery in Hamburg into an office.

To fit the contemporary needs of the modern office block, it was necessary to gut the entire office space, resulting in a confrontation with the stimulating synthesis between the historical façades and the specific needs of the interior office space. The decision was taken to develop a unified concept for the whole building whereby the materials, surface areas and colours would harmonise together.

德国Spine设计事务所完成一个设计，将汉堡的一个啤酒厂改造成了办公室。

London designers Sybarite have completed the interior of a store in Frankfurt for Italian knitwear brand Stefanel.

The components, mainly gloss lacquered GRP and polished stainless steel, are fabricated off site and come together fluidly to form the walls, display surfaces and furniture, creating the impression of a continuous form moulded into display props and surfaces. The sense of composition is reinforced by the attention paid to the smallest detail. （http://www.sybarite-uk.com）

伦敦设计师Sybarite完成了意大利针织品品牌Stefanel位于法兰克福的店面室内设计。

New York and Beijing designers Elevation Workshop have completed the interior of a clothing store in Beijing, China.

The "L-Container" functions dually as a spacial volume and a separation element. The floor inside was lifted to represent a stage. Within the pre-existing conditions, including limited ceiling height and openings, the show room is to focus on the transformation and transition of space through subtle and responsive material changes.(http://www.elevationworkshop.com/)

纽约及北京设计公司ELEV设计事务所完成了一个位于北京的服装店设计。

Norwegian designers Ralston & Bau have completed the interior of a restaurant in Paris.

Both cultures have influenced the theatrical interior concept. The classic opera and geisha cultures from France and Japan can be seen in the scenographic layers, costumes and shadows cast throughout the interior. A contrast of dark and light spaces divides the room: the bright area with the bar as a centre point is used for cooking courses during the daytime, and the dark part, following a perspective angle, including a VIP space to enjoy a gastronomic menu at night.

挪威设计师Ralston & Bau完成了一个位于巴黎的餐厅设计。

Ukrainian architects 2-B-2 have designed a conceptual shop or exhibition stand.

Each plane is a separate graphic composition, while all the planes form a spatial composition. The viewer moves in space, and the relative positioning of elements of a composition varies with each step. The graphic passes in volume. It is difficult to distinguish a plane from volume in an interior.

乌克兰2-B-2设计公司最新设计的空间模型，可以认为是一个概念的店铺，或者是一个展示。

Morningstar

晨星办公

It was to provide open workspace with semi-private and private support amenities. Mixed meeting areas with work zones, break-outs and café areas were to encourage chance meetings with generic workplace modules to allow for innovative equal work methods, flexibility and open communication. Above all the outcome was to be viable, an enduring design solution allowing long-term benefits for Morningstar's growth and its staff. Morningstar sought a classically designed space focusing on restraint, functionality and longevity. The design wasn't to be about revolutionising design principles; instead Morningstar required the fit out to represent their non-hierarchal relaxed professional brand and its position in the financial market. Morningstar required a work environment that focused on open communication that would assist the integration of the merged companies.

It is a restrained, warm, welcoming and comfortable space in its feel, yet is a space that provokes people to see the company professionally much like the people who work there. It is a conservatively luxurious environment focusing on the architectural elements of Harry Sedlier's masterpiece Australia Square.

Being on the highrise levels of Australia Square there needed to be a level of transparency so all access the sweeping Sydney views. The support areas utilise the existing core of the building, allowing the realisation of the transparent level. The interconnecting stair is positioned so the views are maintained while allowing direct access for clients and staff via the café and conference rooms.

The material palette is elegant, classic and rich to portray Morningstar in a sophisticated and professional manner. The detailing is honest, refined and understated, enhancing the beauty of the selected materials.

开放式的工作区内带有半私密和全封闭的小空间；工作区、休息区和餐厅都可用于员工之间的交流——这一具体想法具有可行性。晨星一直认为创新并不是要彻底摒弃已有的设计准则，他们需要一个能够打破等级制度、体现公司金融市场地位的放松工作氛围。

设计师以打造简约、舒适、温馨以及专业性十足的空间氛围为理念。低调奢华的空间环境注重突出各种建筑元素。

鉴于建筑的高度，必须充分利用透明元素，借以将悉尼的迷人景色"邀请"进来。楼梯相互交错存在，便于员工和客户直接从餐厅和会议区走进办公区。

设计中选用的材料典雅而丰富，在专业、精致的手法诠释下，更增添了办公空间的魅力。同时，精细而质朴的细节更是突出了材料本身的美感。

LOCATION
Sydney, Australia
DESIGNERS
The World is round Pty Ltd
COMPLETION
2007
PHOTOGRAPHERS
Tyrone Branigan

项目地点
澳大利亚·悉尼
设计师
圆圆世界设计公司
完工时间
2007
摄影师
蒂龙·布兰尼甘

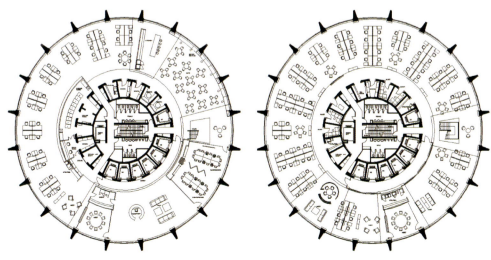

Engine Offices

Engine通讯集团

LOCATION
London, UK
DESIGNERS
Simon Jordan
COMPLETION
2008
PHOTOGRAPHERS

项目地点
英国,伦敦
设计师
西蒙·约旦
完工时间
2008
摄影师

Jump Studios has designed an animated but suitably slick new office for London-based communications group Engine.

Part of Engine's appeal is that clients can opt for multilateral or unilateral engagement with the various partner companies, which meant that Jump Studios had to provide for group working as well as satisfying each of the companies' individual requirements. The move to the new building on Great Portland Street brings all 12 businesses together (across five storeys) for the first time, but while the working floors were kept reasonably generic to allow for the personalisation of space, the building as a whole is characteristically Engine. "The communal areas are where clients and staff are confronted with the Engine group as a single entity and this is where had to reflect what we called the Engine DNA," says project architect Markus Nonn.

The team worked with the concept of "precision engineering". This idea is most clearly manifest in a series of perfectly formed elements that run through the building, essentially forming a backbone that links the ground floor to the fifth. One of the most dramatic is the floating auditorium at entrance level.

Among many other "talking points" in the building are the seating pods on the fifth floor with their Corian shells and Barrisol light ceilings. Here employees are encouraged to interact, serviced by a café offering spectacular views across the city's rooftops and a series of conference and meeting rooms ranging in design, size and style. Nonn explains: "We wanted to offer a more refreshing alternative to the conventional table and chair." Some of the more imaginative solutions include "mini auditorium" seating systems and one room clad entirely in cork (and with cork stools to match) allowing for quick, non-permanent customisation.

Jump工作室为伦敦Engine通讯集团设计了一个全新的、活力十足的办公空间。

集团的要求是每个公司可以多方面或单方面的参与到兄弟公司的设计中来,这就意味着Jump工作室要提供小组工作的模式,同时满足每一家公司的需求。Engine集团搬迁到大波特兰街一幢新建筑中,12家公司共占据五层空间。在Engine的大环境下,每家又拥有其各自的不同风格。"公共区无论对于客户还是员工来说都代表了Engine的主旋律,或者说这里是Engine的核心。"项目建筑师马库斯·诺恩解释说。

设计团队一直秉承着"精密工程"理念,这一理念在一系列建筑元素中清晰体现,如脊柱(连接一层到五层的主要结构)的设计。入口一层浮动的礼堂用于演讲等活动,极为引人注目。

此外,五层的休息亭亦格外吸引眼球,可丽耐板外壳以及软膜天花板特色十足,员工在这样的环境中变得更加善于交流。坐在小咖啡馆里欣赏城市的壮丽景色,也堪称一道特色。风格、规格各异的会议室更是打破传统的桌、椅布局,给人别样的新鲜感受。这一建筑中还有很多极具创意的设计,如迷你礼堂的座位系统、软木材料打造的房间等等。

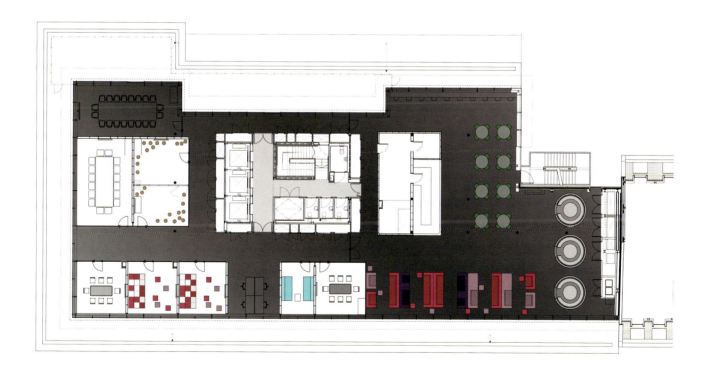

Nike Offices and Showrooms

耐克办公及展厅

Situated over three floors and totaling 1500 square metres, the office houses the marketing, advertising, PR, sales, design functions and products showrooms where key retails are presented with the latest Nike products.

The brief was to develop an inspirational and upbeat working environment reflective of Nike's dedication to performance through innovation and design. This is a temple to kinetics, where visitors are drawn, almost passively, into an unstoppable dynamic.

As a strategically important global hub with a constant flow of visiting personnel from around the world, the London offices and showrooms have been designed to offer a flexible work environment with three quarters of the floor area devoted to meeting spaces. A variety of innovative environments have been designed to accommodate a wide spectrum of meeting opportunities from small huddle spaces through to open café areas, enclosed informal meeting rooms, creative meeting spaces and formal conference areas.

耐克办公及展厅占据三层共1500平方米的空间，包括市场部、广告部、公关部、销售部、设计部以及产品展厅（展示耐克最新的产品）。

客户要求打造一个激发灵感、乐观向上的办公氛围，通过创新和设计展示耐克的理念。据此，设计师创造了一个动感十足的空间，来客情不自禁地被吸引。

作为该品牌的国际枢纽站，伦敦公司每年都会接待世界各地的来客，因此设计师将四分之三的办公空间打造为会议区。无论是簇拥一团的小空间、正式会议室还是开放式的餐厅，都可用作各种集会。

LOCATION
London, UK
DESIGNERS
Jump Studios
COMPLETION
2008
PHOTOGRAPHERS
Mark York

项目地点
英国.伦敦
设计师
Jump工作室
完工时间
2008
摄影师
马克·约克

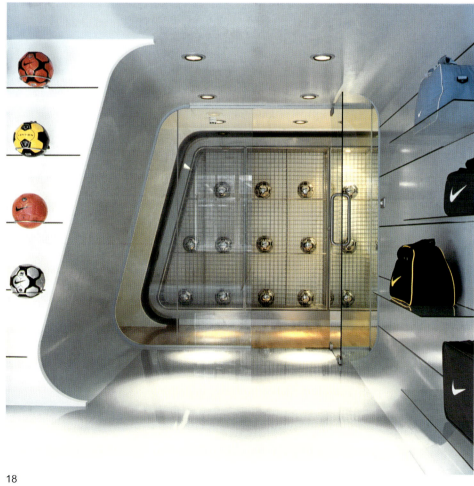

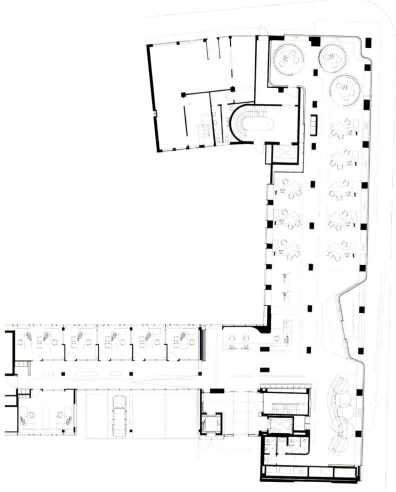

Offices in Torre Murano

托莱·穆拉诺办公

This client needed to move to a more corporate space without losing that warm feeling they had achieved in their offices along time.

The space is divided into two areas, the technical part and the corporate side, joined by a zone formed by formal and informal meeting spaces. This is a transition that provides service to the whole office without doubling square footage.

The colour selection and combination was an essential part in the design because the designers try to reflect the colours of nature to recreate the exterior feeling in the interior space. The use of blues, greens and oranges together helps to increase productivity and purify the ambience.

It is about a peaceful space with few private offices, a big open workspace and lots of alternative work stations, about the changes the company is making in how they work. New generations are taking the company towards the future so the space was designed to be dynamic and modern to comply with the demanding needs.

A sober, yet colourful, reception welcomes the office, allowing the user to glance at the phone booths behind the threshold. Private offices and meeting rooms were located in the south façades in order to help reduce heat and help lower energetic costs. In addition, this is an office designed to be as sustainable as possible, thought to be an ethical space for everyone.

客户需将公司搬迁到一个更具商业氛围的空间，并保留他们长期以来营造的温馨氛围。新的办公空间被分为两个区域——技术区和办公区，中间通过正式或非正式会议区连接，功能全面而又节省空间。

色彩选择和搭配是整个设计的重要组成部分，设计师尽量使用自然的颜色以便将室外的氛围"移置"到室内。蓝色、绿色以及橙色的混合使用在净化空间的同时，更能激发员工的创造力进而提高工作效率。

几间单独的办公室、一个开阔的开放式办公区以及一系列的工作台突出了公司模式的变化。年轻的一代是公司的主力，他们引领公司前进，因此为满足他们的需要，设计师刻意营造了动态感和现代感。

接待台淡雅而不单调，来客在此可以清晰地瞥见门槛后面的电话亭；单独的办公空间及会议室沿着南面一侧设置，减少热能损失及降低成本。此外，设计中最大程度体现可持续发展理念。

LOCATION
Mexico City, Mexico
DESIGNERS
Space
COMPLETION
2008
PHOTOGRAPHERS
Space

项目地点
墨西哥.墨西哥城
设计师
Space建筑师事务所
完工时间
2008
摄影师
Space建筑师事务所

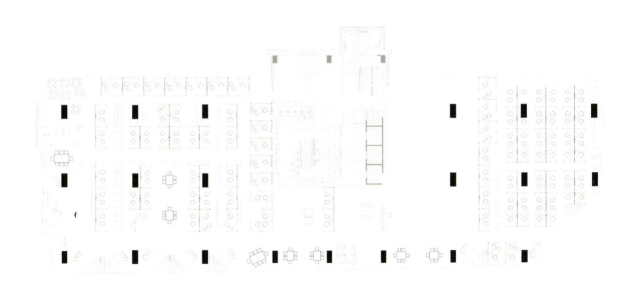

Alsea

艾尔西公司

SPACE worked hand in hand with Alsea's directors to define a concept that responded to their needs and which, at the same time, suggested an image consistent with the brands.

The resulting concept was the idea of generating "Alsea City", a space with an urban touch that operated on the basis of neighborhoods, each neighborhood bringing together a brand, and each of these spaces having a "neighborhood centre". The neighborhood centres became the ideal pretexts to generate identity to each group through informal meeting areas or what they have come to call "casual collisions".

The architectural layout arises from the idea of maximising natural light, distributing the private offices perpendicular to the façades, thus liberating the façades to the highest possible extent. The number and form of the board meeting areas were defined after an extensive investigation. Some of these areas were resolved with unconventional spaces such as pool tables and informal lounges.

The space is formed by two architectural storeys intercommunicated by an interconnecting stair and, in the centre of these spaces, a mezzanine that works as council room was designed. This council room came into being through a glass box lined with texts that express the company's culture and values.

Moreover, this space has categorically helped to transform the way of organisation; it contributes to a more dynamic communication fostering the company's culture, and has at the same time become an icon of corporate design in Mexico.

应客户要求，设计师同艾尔西总监共同合作，打造了"注重互动性及动态感、彰显公司'年轻、魄力十足'企业文化"的设计理念。

这一理念最终通过"营造艾尔西城"的构想实现。带有城市韵味的"空间"由不同的"社区组成"，每个"社区"代表一个品牌，社区中心则作为自由交流区。

两层的办公空间通过楼梯连接，中层楼如同一个"玻璃盒子"，四周写有介绍公司文化和价值的文字，可用作会议室。

设计师在构建空间格局时以"最大程度地利用自然光线"为理念，私人办公空间与墙面垂直设置便于墙面延伸。会议区的数量及形状都是通过大量的调查之后设定的，其中一些区域的布置打破传统，放置了桌球台或长沙发。

需要指出的是，这一创新的空间设计模式改变了传统的办公结构，突出互动性和动态感的同时，强调企业文化。如今，这一设计模式已成为墨西哥地区办公空间设计的模板。

LOCATION
Mexico City, Mexico
DESIGNERS
Space
COMPLETION
2007
PHOTOGRAPHERS
Santiago Barreiro, Pim Schalkwijk

项目地点
墨西哥.墨西哥城
设计师
空间设计建筑工作室
完工时间
2007
摄影师
圣地亚哥·巴雷罗.皮姆·肖吉克

Lista Office

丽思塔办公

The room glares in white! The white ribbons entangled through an aluminium construction create an organic light-flooded tunnel in order to antagonise the tubular-like basement situation. The designers intended to lighten the appearence of an exhibition system by using a flexible and movable material. The elements also seperate the room and guide the visitor through the exhibition without minimising the space.

By extending the ribbons the visitors´ view will expand according to the perspective view. The 1000-meter-long ribbon are tightened dynamically through the room from base to ceiling in a special adjustment. According to the visitors´ view, shadows generate graphic elements on the ground next to moire effects within the ribbon-walls appearing through the whole exhibition-stand area.

The exhibited company not only shows the newest objects of its collection but also presents objects of the first newcomer-award brought to live by Lista office. The best 10 designs are shown in a seperate space. Visitors are pleased to glance at the products by pulling the ribbons apart and have a closer look. In order to underline this special show, the ribbons are in shiny black and the awarded objects are tied to the ceiling, giving the scenery its more spherical look.

整个办公空间沉浸在纯净的白色之中——白色丝带"缠绕"在铝制结构上，形成了一个光线满溢的"隧道"，赋予管状的地下室空间截然不同的氛围。设计师采用灵活而便于移动的材质，旨在点亮整个展示系统。同时，这些元素将空间分割，并指引着来访者在展区内穿行。

随着丝带的延伸，参观者的视野随之拓宽。1000米长的丝带从地面蜿蜒攀升到屋顶，动态感十足。在参观者看来，丝带的影子投射到地面上，形成了平面元素，为整个展示区平添一丝活力。

公司的展区内不仅包括其最新的产品，同时还有新来员工的首个作品，其中"十大最佳设计作品"被单独展示。参观者将丝带掀开，便可清晰地看到里面的产品。此外，为突出特殊的展品，丝带采用亮黑色装饰，获奖作品悬垂在屋顶上，营造了一种漂浮感。

LOCATION
St.gallen, Switzerland
DESIGNERS
Carmen and Urs Greutmann, greutmann bolzern designstudio
COMPLETION
2008
PHOTOGRAPHERS
Claudia Below, Greutmann

项目地点
瑞士.圣加伦
设计师
Greutmann Bolzern设计工作室
完工时间
2008
摄影师
克劳迪娅·比露, 格鲁特曼

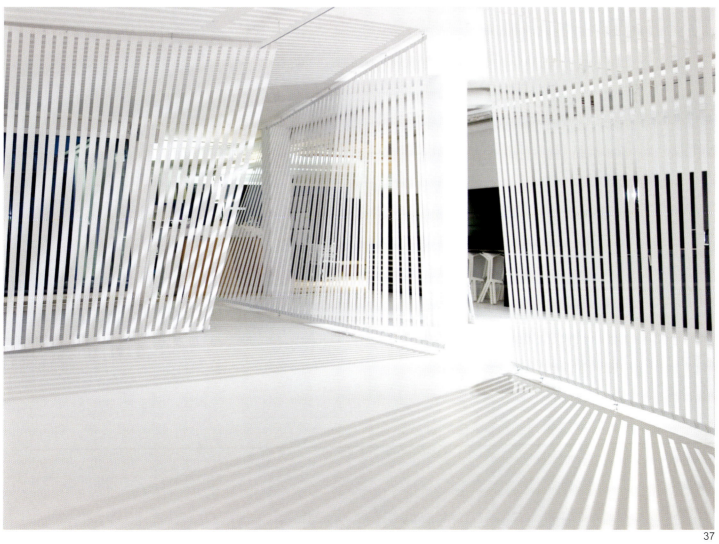

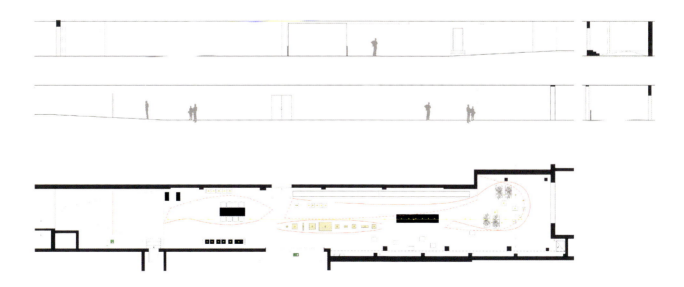

LG European Design Certre

LG欧洲设计中心

Jump Studios has recently completed the designs for a new European Design HQ for LG, a leader in consumer electronics and mobile communications. Relocating from Milan, Italy to Covent Garden, London, the European Design Studio of LG is one of the most important design centres for the electronic goods company and houses a team of 20 dedicated industrial and interface designers, drawn from around Europe. They will design products for LG's entire range of consumer goods including mobile handsets and devices, flat-screen TVs and audio systems, white-goods and other home electronics.

"Our new European Design Centre has been established to provide our design team with the optimum creative environment, which includes extensive libraries and leading technologies to aid in the design process. There are significant cultural and technical differences between Europe and other markets. As a result, for LG to grow its market presence in Europe, it must invest in and be sensitive to the consumer's requirements, while responding to technical and cultural differences," said Mr James Kim, president and CEO of LG Electronics Europe.

LG——电子以及移动通讯产品的领航者，已将其欧洲产品研发设计中心从米兰搬迁到伦敦的中心广场区。作为LG电子产品公司的研发供应商，中心共有20名来自欧洲不同地区的专业设计师（工业及界面设计）。产品包括移动电话及配件、平板电视、音频设备、大型家用电器等。

詹姆斯·基姆先生——LG执行总裁曾经说过："创立新的欧洲研发中心的目的就是我们的设计团队提供一个适于创作的环境，包括宽敞的图书馆以及先进的设备。考虑到欧洲市场同其他地区的文化及技术差异，LG必须全面感知客户的需求，实现市场价值的增长。"

LOCATION
London, UK
DESIGNERS
Jump Studios
COMPLETION
2008
PHOTOGRAPHERS
Gareth Gardner

项目地点
英国.伦敦
设计师
Jump建筑工作室
完工时间
2008
摄影师
加雷斯·加纳德

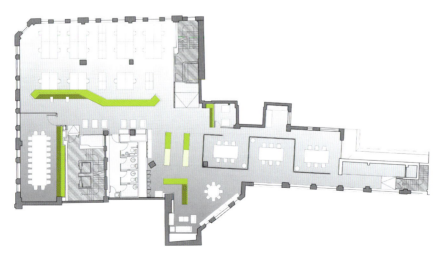

CheBanca!

CheBanca! 银行

The design of the new branches emphasises the consultancy, self transaction and self education activities, bringing the consumer at the centre of the process.

The Natural Tech concept designed by Crea International is inspired by "the rules of simplicity" by John Maeda. The leading design idea is that the things that surround us have to get back to essential. The overall atmosphere of CheBanca! recalls the warmth and light of the sun, and the layout of CheBanca! reminds the logical organisation of the solar system with the client ideally at the centre of it. The Natural Tech of CheBanca! means ethic and transparency of a world that does not deceive and the apparently "technology free" environment brings human contact upfront.

The yellow colour that permeates the environment reminds of the sunshine light; the aniline-treated wood suggests a straight forward approach; the metacrylic material printed with the honeycomb texture casts a friendly atmosphere.

Key innovations in this project are various. First of all, the layout overturns with the presence of the central base point and the perimeter connection booths. The second important point is the environmental branding specifically focused on strong visual elements such as the portal, the windows, the interactive walls that transfer the product offer and the info about the cultural activities of the local community and two relax areas, coffee bar and kids garden.

The CheBanca! concept is completed with a mobile branch conceived as communication tool to complete the traditional media offering the opportunity to get in touch with new potential clients.

Chebanca！银行支行的设计强调咨询、自行交易和自我教育等活动，是消费者成为整个环节的中心。

克雷亚公司采用的设计理念就是"自然技术"。主要的设计想法是——我们周围的一切事物最终都会回归到本质：太阳的光和热，这是一个科技"太阳"。在银行中，人们都被这个特殊的科技"太阳"的光线笼罩着。银行的布局也能让人们联想到太阳系的逻辑结构，而客户在其中心。银行采用的自然科技就是想表达一个没有谎言、纯洁、道德的世界。

整个空间"弥漫"着黄色，让人不禁联想到灿烂的太阳光线；经苯胺处理的木材展现了自然、简洁的设计方法；带有孔状结构的丙烯材料营造了温馨、友好的氛围。

其中主要的创新特色包括两个方面。首先是空间格局的变化。原有的格局被打破，形成了中心区和四周的服务区。其次是视觉元素，如入口、窗户、互动墙壁上标识的设计。

移动式支行的设计更加完善了CheBanca!的整体设计理念，可以被看作是一种通讯工具，实现银行与潜在客户之间的联系。

LOCATION
Milan, Italy
DESIGNERS
Crea International s.r.l.
COMPLETION
2008
PHOTOGRAPHERS
Beppe Raso

项目地点
意大利·米兰
设计师
克雷亚国际设计公司
完工时间
2008
摄影师
比派·拉索

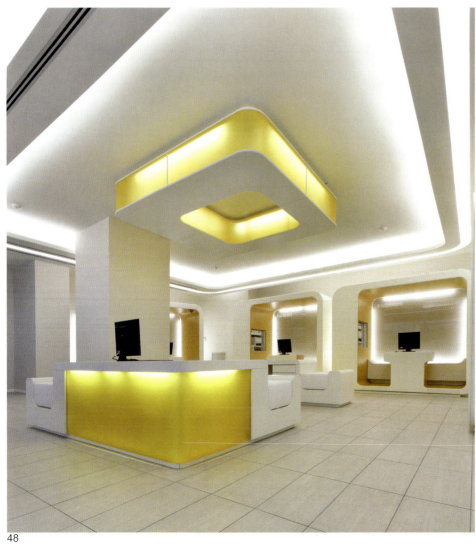

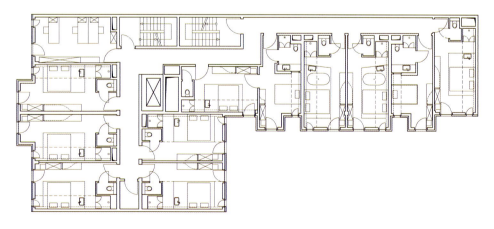

New Google Meeting Room

谷歌新会议室

A key element in the design approach was that the Zooglers should participate in the design process to create their own local identity.

The architects began with a rapid process of research and analysis to map out the opportunities and challenges posed by the building as well as the emotional and practical requirements of the Zooglers. The latter was achieved by conducting a survey of all the Zooglers, complemented by a series of workshops and interviews.

Although the details of the survey outcomes remain confidential, the process revealed that the optimal working environment for Zooglers needed to be diverse and at the same time harmonious whilst making it a fun and an enjoyable place to work in. The survey also showed that while personal workspace needed to be functional and more neutral, communal areas had to offer strong visual and more aesthetically enjoyable and entertaining qualities to stimulate creativity, innovation and collaboration.

The Zooglers decided early on that they preferred to reduce their personal net area of workspace in order to gain more communal and meeting areas. The working areas were therefore designed with a high degree of space efficiency. Additionally, they had to be able to accommodate frequent staff rotation and growth. On average a Googler moves twice a year within the building, consequently the office layout was designed for maximum adaptability so that all groups and departments can use any part of the office space. The office areas are organised along a central core and are a mixture of open-plan workspaces for 6-10 people and enclosed offices for 4-6 people.

设计中的关键因素是请谷歌的员工参与设计过程，建立属于自己的特色。

设计师迅速地研究和分析了建筑中的优势和缺陷，以及员工们的实际需求，并对所有员工进行调查，开展了一系列的讲习班和演讨会，了解他们的需求。

调查结果是保密的，但这一过程表明，谷歌的员工需要多样化的工作环境，同时还要富于趣味性，令人心情愉快。调查显示，员工的工作场所强调功能性，个性化色彩不太强烈；而公共区域则要有强大的视觉冲击力，增加美的享受，还要有娱乐性，以促进创造、创新和协作。

员工们决定减少自己办公室的面积，以增加公共区和会议室的面积。因此办公室的设计充分利用了空间。另外，办公室还要能适应频繁人事变动和人员增长。谷歌员工平均一年要更换两次办公室，因此办公室布局的设计要具有强大的适应性，让各种人群和不同部门在这里工作起来得心应手。办公区设在各楼层的中心地带，开放式混合办公室能容纳6-10人，封闭式办公室能容纳4-6人。

LOCATION
, Switzerland
DESIGNERS
Camenzind Evolution Ltd.
COMPLETION
2008
PHOTOGRAPHERS
Camenzind Evolution Ltd.

项目地点
瑞士.
设计师
卡门辛德发展公司
完工时间
2008
摄影师
卡门辛德发展公司

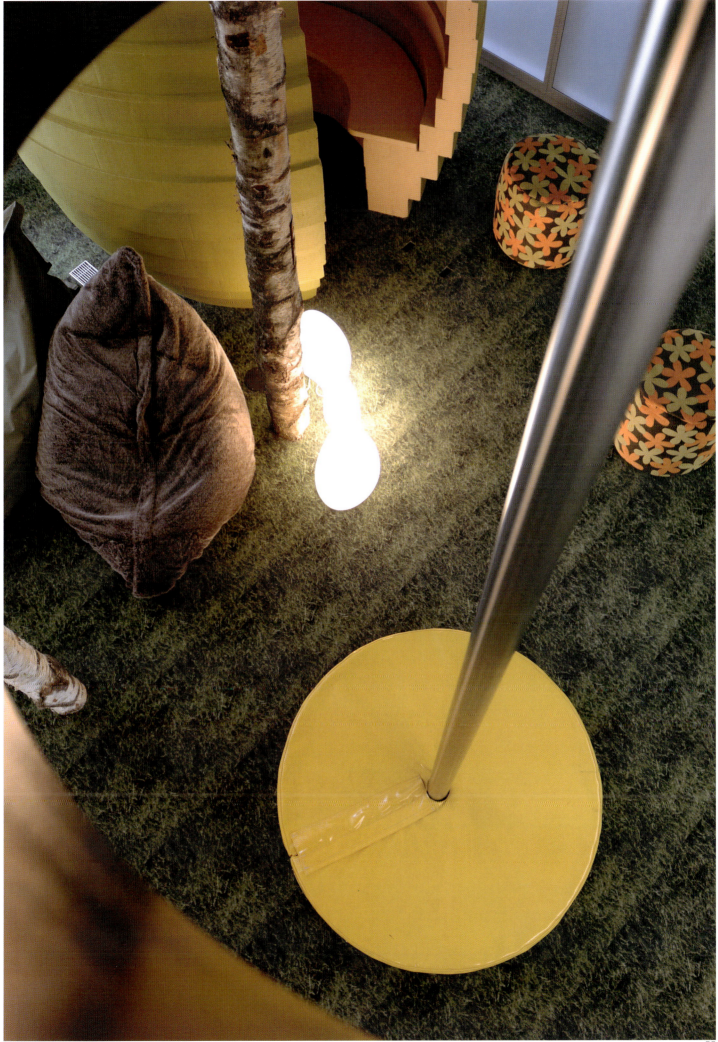

Banco Deuno

Deuno银行

Banco Deuno (Your-own Bank)confided in usoarquitectura to transform all the institutional precepts and corporate image into an interior design where the main core is the client's experience.

Colour is the central element of the design concept. All the elements, from the collaborators uniforms to the front pieces integrated to the different buildings where the branches are located, are part of the daring colour palette. The spatial language generated by this selection of colours, not common for bank institutions in Mexico, contains all the clients' needs stipulated in the programme.

Every branch is different from the other representing a very big challenge for usoarquitectura who translated it into the integration of the general concept for each one. The different spaces are standardised by the correct selection of finishes and colour application, being this last one the personal signature of each branch. The bank's dynamic culture is interpreted through this ample colour palette. The combinations are flexible and can be adapted according to each space needs. Architects Gabriel Salazar and Fernando Castañón applied their knowledge and experience in commercial interior architecture developing adequate areas for both the internal operation of the bank and customer service.

Each branch is different but everyone has the same shape and colour language. No matter which branch the client is in, he will feel in Banco Deuno. The communication campaign and the new bank concept were also translated in a series of text and image messages that combined with the colour palette complement the dynamic of the bank.

Deuno（自己的银行）邀请USO建筑工作室为其进行室内设计，并将公司体制规章及企业形象融入其中，注重客户的体验。

色彩是整个设计理念的核心元素。从员工制服到支行门前挂置的牌匾全部采用大胆的颜色设计，这在墨西哥银行机构中是极为罕见的。设计师通过选择色彩实现空间建筑语言，并借此满足客户的所有需求。

所有的支行采用统一的设计理念，但每一家又别具特色，这对于设计师来说未尝不是一项艰巨的任务。不同的空间通过色彩和装饰物的选择达到统一的效果，进而体现银行动态的文化底蕴。色彩可以灵活搭配，满足不同的需求，同时构成各家支行的特色。建筑师将他们的知识和体验淋漓尽致地渗透到商业空间的室内设计上，为客户和银行内部运作打造了足够的空间。

虽然每家支行都具备其自身的特色，但他们具备相同的造型和色彩语言，因此无论客户走进哪家都会感到Deuno的风格。此外，银行的信息活动和全新设计理念通过文字和图片展现出来，更增添了银行的动态性。

LOCATION
Mexico City, Mexico
DESIGNERS
usoarquitectura/ Gabriel Salazar y Fernando Castañón
COMPLETION
2007
PHOTOGRAPHERS
Tygre

项目地点
墨西哥,墨西哥城
设计师
uso建筑工作室
完工时间
2007
摄影师
Tygre

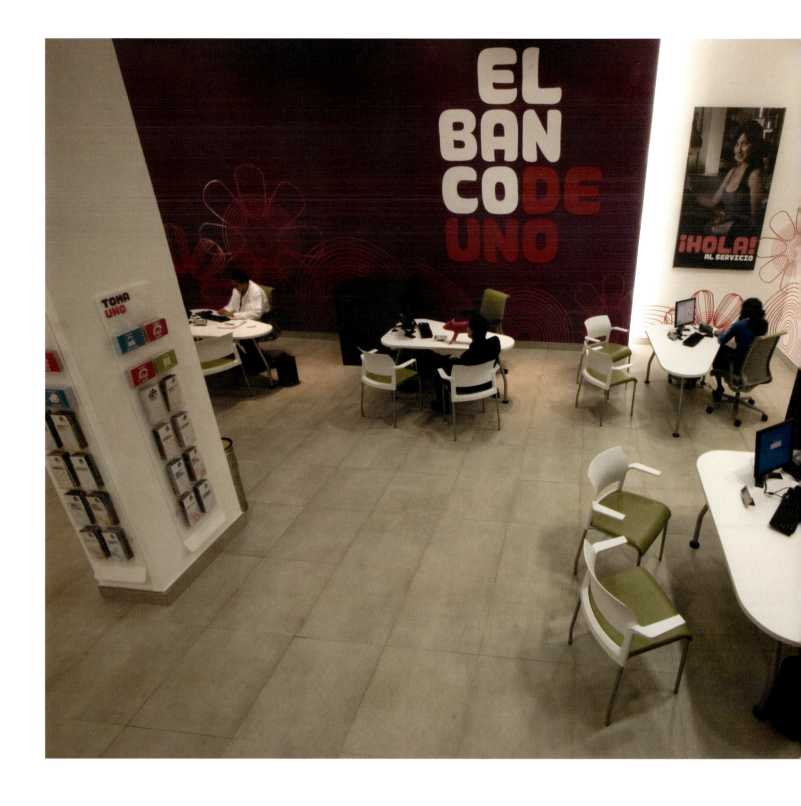

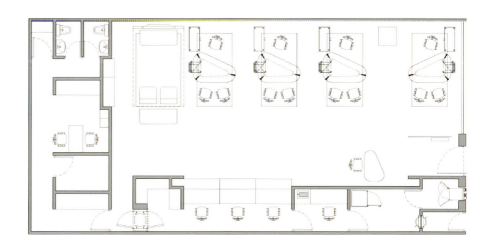

Mercat a la Planxa

莫卡特餐厅

Mercat redefines the "hotel restaurant". The concept was to create a destination experience that targeted the city of Chicago, not the hotel guest. In addition, the sequence of space is unique (if not challenging), with a ground floor entrance and a second level restaurant. The solution creates a unique sequence of interior spaces that allow a choice of experience for the customer. The design does not romanticise Spanish design; instead, it captures the vibrancy of modern Barcelona.

For Mercat, the designers drew from modern Spanish aesthetics to create a vibrant ambience reminiscent of Barcelona's bustling markets. The entrance to the restaurant is through a small bar on Michigan Avenue. The guest enters this intimate space, clad in walnut, to evoke the feeling of a classic Spanish bar. The back of the bar is surrounded by a curved screen composed of multi-coloured hex blocks that is also the back of the large circular stair that leads to the second floor dining room.

The dining room is a grand two-storey space. The designers created a layout which radiates from the space so that all three levels of the main restaurant allow guests to have prime views through the giant windows that overlook Grant Park. The space features a large open kitchen, a winding central staircase and a mezzanine with a private dining room. A three-storey sculptural screen, upholstery with fuchsia and vibrant green accents, and natural porcelain slab floors, contrast the restaurant's modern elements against the hotel's classic edifice.

莫卡特餐厅改写了"酒店餐厅"的定义，旨在吸引芝加哥食客，而非酒店住客。除此之外，餐厅在格局布置上也别出心裁，其设计不仅仅彰显西班牙风格的浪漫色彩，同时突出了巴塞罗那的现代与活力。

设计师从西班牙现代美学元素中获得灵感，营造了一个活力十足的空间，让人不禁联想到巴塞罗那市区内人潮泉涌的市场。餐厅入口设在密歇根大街上的小酒吧内，从此穿过，客人便可进入宁静而亲切的空间。酒吧后部被弯曲的屏风包围，并一直通往楼上的就餐区。

就餐区占据两层空间，高大的窗户将格兰特公园的美景完全收揽进来。宽阔的开放式厨房、蜿蜒的中央楼梯以及带有私人餐区的包厢构成了餐厅的特色。紫红色和绿色相间的雕塑屏风、简约自然的陶瓷地板与酒店古典的外表形成鲜明对比。

LOCATION
Chicago, USA
DESIGNERS
d-ash design
COMPLETION
2008
PHOTOGRAPHERS
Frank Oudeman

项目地点
美国,芝加哥
设计师
德－阿什设计公司
完工时间
2008
摄影师
弗兰克·欧德曼

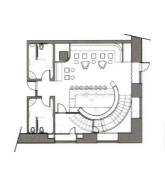
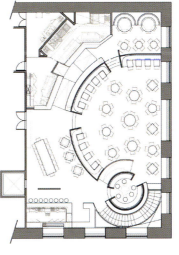
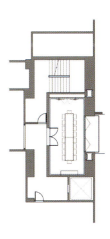

Amici Mi

艾米西餐厅

Downtown Kiev has seen the opening of a new restaurant, which has found its new content in the catering context of the city. The overall aesthetics of the restaurant's interior is subordinated to the priority of food and visitors, which is perceived as a kind of laconic neutral passe-partout for the repast. The project features luxury clothed in the gown of democracy, refinement disguised in sophisticated simplicity – all this being distinctive features of modernism.

In fact, there were no purely architectural tasks involved here: the shell structure was set by the layout of the building (the restaurant occupying the ground floor). The authors' intention was to neutralise the existing pylons, avoiding their involvement in spatial zoning. The ceiling is also perceived as a neutral element. The overall idea behind that was to interpret the walls, the ceiling and the floor as framing surfaces with the emphasis on what is happening inside.

The sofas were manufactured by detailed author's drawings. As for the other objects, the authors first shaped out the desired forms and then, on looking through catalogues from supplyers, they made selections, preserving maximum silmilarity with the previous sketches.

One of the most important scenarios was to create a space of iceberg-isles, both holistic and discrete. A search for a space and a structure for circulation by methods of urban planning, modulating quarters and streets, has resulted in an environment for privacy, visual contacts, and maybe flirting.

艾米西餐厅的新建更加丰富了基辅城的饮食文化，其室内设计以突出食物和来客为主，简约而不乏精致。

餐厅占据整个一层，设计师将原有的塔门加以修饰，以避免打乱空间的节奏。墙壁、天花板以及地面作为框架结构，突显中性风格，设计的重中之重便落到了框架所围合的"内部结构"上。

沙发是严格按照设计师的手绘图打造的，其它用品则是设计师首先勾勒出大概图样，然后同供应商仔细协商之后选择的，当然要尽量与最初的概念相一致。

值得一提的是设计师运用城市规划的方法在餐厅内营造了一个私密而并不闭塞的环境。

LOCATION
Kiev, Ukraine
DESIGNERS
Oleg Drozdov, Alexander Zhuk
COMPLETION
2007
PHOTOGRAPHERS
Andrey Avdeenko

项目地点
乌克兰.基辅
设计师
奥莱格·德罗兹多夫.亚历山大·朱克
完工时间
2007
摄影师
安德烈·艾夫丁科

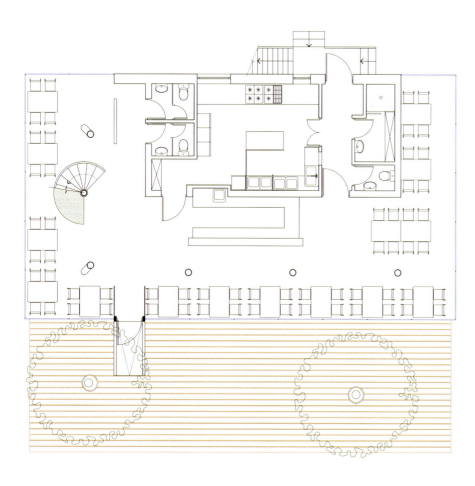

Hamasei Restaurant

哈马赛餐厅

The extension plan of the restaurant, developed by the architecture practice Alvisi Kirimoto + Partners, concerns a surface of 250 square metres integrated to the existing 300 square metres, maintaining a strong Japanese connotation with a slight Western influence and a contemporary mood.

Hamasei sushi-restaurant offers the public a sophisticated environment, warm and pleasant at the same time, where the use of wood and earth colours is combined to a careful dosage. The light, specifically adjusted on every table like an essential architectonic element, confers an intimate atmosphere to the spaces. The strong character of the project is emphasised by some important elements like the Japanese calligraphic work in the first room. The symbol, printed on the backlit paper, creates the background for the tables and of the white sofa running all along the wall, a further accent added to the space.

Wooden panels, installed to the walls and ceilings leaving a small distance, which transmit the idea of a decomposable box, express the Japanese concept of modularity, applied both in architecture and philosophy.

The architects have studied every architectonic element and selected or specifically designed the furniture, in order to guarantee the general visual soberness by the accuracy of each detail and of the technological solutions and plant engineering, leaving to the guests the freedom to relate or not to other people.

Hamasei project by Alvisi Kirimoto + Partners will proceed with the renovation of the restaurant's first part expected.

设计师的任务即对原有的餐厅进行扩建（餐厅原有面积300平方米，新增250平方米），突出浓厚日式韵味的同时兼顾西式风格及现代特色。

设计师精心地将木色与土色混合使用，营造了雅致、温暖而又舒适的环境。灯光作为一种建筑元素被使用，每张餐桌上的光线都加以精心调试，增添了空间的亲切感。日式的书法作品以及象征物打造了意蕴独特的背景。

墙壁及天花板上的装饰木板以一定间距排列，如同一个个可分解的盒子，展示了日式建筑学及哲学中的模块理念。

此外，设计师仔细研究每个建筑元素、精心挑选每件家具，注重突出细节，创造视觉美感。

餐厅的进一步扩建工作将继续由Alvisi Kirimoto合伙人事务所完成。

LOCATION
Rome, Italy
DESIGNERS
Alvisi Kirimoto + Partners
COMPLETION
2007
PHOTOGRAPHERS
Luigi Filetici

项目地点
意大利.罗马
设计师
Alvisi Kirimoto合伙人事务所
完工时间
2007
摄影师
鲁吉·菲莱提斯

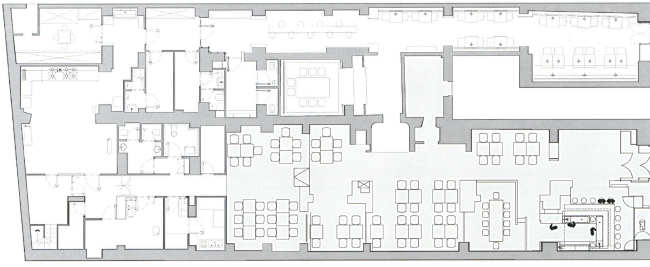

Jaso
Restaurant

雅索餐厅

Serrano Monjaraz Arquitectos were asked to develop the interior design for a new restaurant. Step by step the concept was designed based on the sensorial elements: fire, air, water and sound. The materials play a very important role in this project. For example, wood, marble and iron are mixed in different percentages and combinations in search for a very natural atmosphere. The clients are greeted at the reception with a big tree trunk that was rescued and incorporated into the layout.

Being a former house from the late 1940s, the project had to be adapted to fit this two-level construction. In the main floor there is a lounge and bar area used both for reception and for the clients that only stop for drinks. Then, a big double-height salon is founded and ends in a fantastic interior garden. This ample dinning room has a very warm atmosphere created by the combination of zebra wood-covering walls and mahogany wood on the ceiling. For the floors a very interesting combination of stone and wood was achieved.

At the back of the main floor the kitchen is located. A metal plated and glass staircase leads to the second level. There is a private salon inside of the cellar and an area designed for private events. These areas end in a spectacular terrace surrounded by flowerpots with an sculpture fountain made of 4,289 pieces of silica sand. The pieces form the columns that are submerged in water and are followed by a rosemary wall.

塞拉诺·蒙雅拉兹建筑事务所受邀打造餐厅室内空间，设计师以"水、火、声、气"等感知元素为理念，将木材、大理石和铁材按不同比例混合，营造朴实自然的氛围。接待台处，高大的树干迎接着宾客的到来。

餐厅分两部分，一部分包括大厅和酒吧，用作接待处和客人小憩畅饮空间；另一部分为主餐厅，斑马木墙板、红木天花板以及石材和木板混合的地面营造了温暖祥和的氛围。

厨房设在大厅和酒吧后方，金属板和玻璃材料的楼梯一直通往楼上；私密就餐区一直延伸到露台处，四周摆放着花盆；4289块硅砂构成的瀑布雕塑结构形成廊柱，沉浸在水中。

LOCATION
México City, Mexico
DESIGNERS
Serrano Monjaraz Arquitectos
COMPLETION
2006
PHOTOGRAPHERS
Pedro Hiriart

项目地点
墨西哥,墨西哥城
设计师
塞拉诺·蒙雅拉兹建筑事务所
完工时间
2006
摄影师
佩得鲁·希里阿特

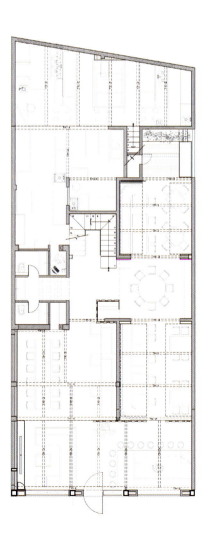

Labels
Clothes Shop

品牌服装店

Steel trees form a reference to the Garden of Eden; white refers to virginal innocence; black is for the lost paradise: a subtle feel for mythology and the mystical is seen in all the work of interior designer Maurice Mentjens. He combined three small spaces and then divided them again into yin and yang, back to the origins of fashion – "the mother of all the arts".

The existing shop soon proved to be too small, and was extended last year with the addition of a neighbouring premises. The long, narrow garden in between the buildings was given a glass roof. By breaking through windows and doors in the side walls, three interconnected spaces were created with seven passageways. Mentjens created a division between the women's and men's sections precisely in the middle of the central glass-roofed space, starting with the floor. One half of the floor is pure white, the other jet black: colours representing yin and yang, the feminine and the masculine. Elegant versus rugged, like the intangible, graceful world of Venus and the earthy, black smithy of Vulcan, her husband, in Roman mythology.

The connection between the spaces is made by two sales counters, half protruding into the glass-roofed space and half into the white or black areas. Thanks to an ingenious anchoring system in the wall, the blocks are almost magically suspended in the spaces. Only the very ends are subtly supported by a plexiglass foot.

钢材打造的树木构成通往伊甸园的标识；白色代表纯洁无暇；黑色象征"迷失的天堂"，设计师莫里斯·门提真斯采用"阴阳"的理念，将以此为主题的三个空间完美融合，并与"时尚即为所有艺术形式之母"的理念相呼应，营造了神话般的购物环境。

2008年，原有的商店被扩建。两幢建筑之间的狭长走廊安装了玻璃顶棚；侧面墙壁上的门窗被拆除；新增的七条过道构筑了三个相互连接的空间。商店中央区域的玻璃顶棚空间被划分成男装区和女装区。其中一部分地面采用纯白粉刷，另一部分则装饰以乌黑色，分别代表着"阴"和"阳"、"女性"和"男性"。典雅与粗糙并存，如同"女神"维纳斯的美好世界与"罗神"（罗马神话中维纳斯的丈夫）粗陋制铁铺。

空间之间通过两个收银台连接，一个延伸到带有玻璃顶棚的区域，另一个通往黑、白区。独特的墙壁固定系统使得隔断巧妙地悬浮在空间内，别具特色。

LOCATION
Sittard, The Netherlands
DESIGNERS
Maurice Mentjens Design, Holtum
COMPLETION
2009
PHOTOGRAPHERS
Leon Abraas

项目地点
荷兰 锡塔德
设计师
莫里斯·门提真斯设计公司
完工时间
2009
摄影师
利昂·阿布拉斯

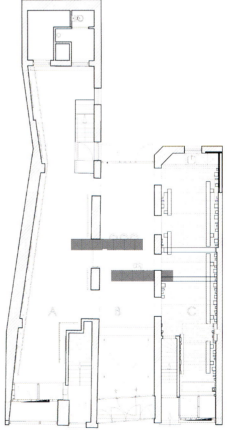

Metal House

金属屋

The "Metal House" concept was selected for the new four-storey development in Taichung City as it was in keeping with the local environment and resonated with the brand's avant-garde industrial image.

To highlight the avant-garde and raw industrial impression, galvanised steel floor slab reinforcements were used for the external façade, an unconventional touch. The ground floor space was treated as a transition between the building and the environment, with the pebble-surface used for the floor to introduce a local flavour. Through the ramp, two types of ascending and descending relationships were created, linking the interior with the external streetscape.

Metal casting provided the inspiration for the sculpted feel of the stairs providing access to the second floor, creating a dynamic system. Curved metallic panels were also used throughout the internal space to create irregular surfaces while heavy slide rails were added to the ceilings to homogenise the visual aesthetics of the overall space.

Mirror surfaces were used for the foreground at each level and wrap around the wall that forms the interface with the outside world. The result is an overlap between the inside and outside that responds to the surrounding environment.

"金属屋"位于台中，是一幢四层新建筑，设计风格融合了当地特色，同时彰显品牌的前卫时尚形象。

设计师选用电镀钢板装饰建筑外立面，营造现代感。一楼地面采用鹅卵石铺设，突显当地特色。坡道将室内与街景连接起来，设计独特。

通往二楼的楼梯采用金属铸件设计，造型独特，为空间增添活力。弯曲的金属板遍布整个室内空间，打造了不规则的表面；天花板上安装了滑道，创造统一的视觉美感。

每层的前厅采用镜面装饰，将室内外空间融合，和谐统一。

LOCATION
Taiwan, China
DESIGNERS
Shichieh Lu
COMPLETION
2008
PHOTOGRAPHERS
Marc Gerritsen

项目地点
中国.台湾
设计师
陆希杰
完工时间
2008
摄影师
马克·格里森

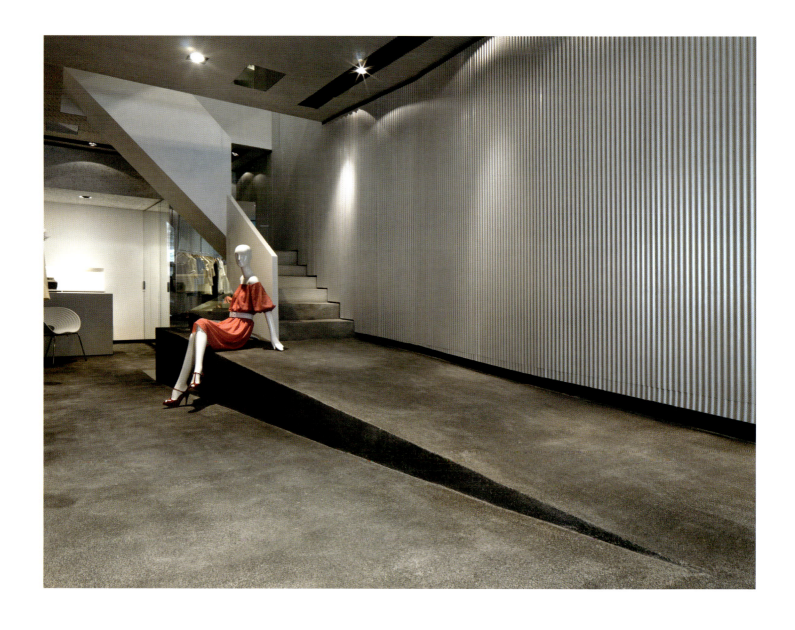

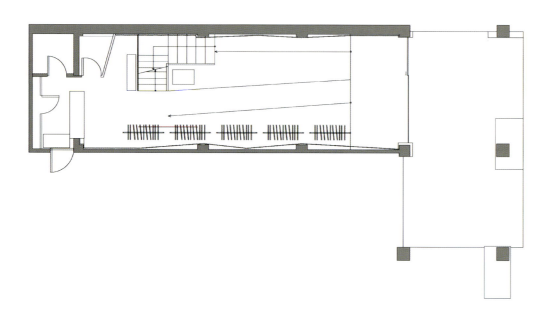

La Grande Aquagirl Shop

水女孩时尚屋

The shop would sell clothes, shoes and accessories, but the brief was to give more importance to the shoes, so designer separate 3 areas and put the shoes area in the centre, on stage.

The heart of the project was to create a surprising environment, thanks to the mirroring effects of the 4 big swiveling mirrors and thanks to the light effect, which seems to come from inside the ceiling or the walls and not from real lighting apparels.Floor, walls, and ceiling are all different nuances of light cool grey, but they look exactly the same.

The light was very important, since the designer decide not to use any spot light but only diffused light. So they use a fluorescent linear lighting hidden in the border of every false ceiling, and square light with compact fluorescent, inside the false ceiling area. The glossy finishing of the floor helps to diffuse the lighting and the mirrors do the same.

Also the big niches have indirect lights hidden in the case and in the top, so to be perceived as bright rectangle in the wall.

All the furnitures of the commercial area are on the design. The cabinet benches and casher are in lacquered shiny white, while all the hanger (self standing and fixed to the wall) are in stainless steel champagne finishing. The hat hanger, conceived as a sculptural piece, was in stainless steel champagne finishing, with fluorescent light in the base.

这间时尚店将出售服装、鞋和配饰，鞋是这里的主要商品，因此设计师划分了三个区域，将鞋子的柜台放在最中央。

设计的重点是打造一个令人惊艳的店面，设计师安装了四面大镜子，并利用灯光效果，看上去灯光像是从天花板和墙壁中照射出来的一样。地板、墙壁和天花板采用了深浅略有不同的浅灰色，但它们看上去却是完全一样的。

灯光十分重要，因此设计师决定不使用射灯，而只用漫射灯。设计师在吊顶天花板的边缘安装了灯带，并在吊顶内侧紧挨灯带的位置安装了方形灯。地板光亮的表面也有助于分散灯光，与镜面的效果相同。壁龛里同样装着间接光源，看上去就像墙上嵌着一个发光的长方形。

在设计中，所有的家具都有商业用途。橱柜式长椅和收银台都被漆成亮白色，所有的衣架（包括站立式的和嵌入墙内的）都是不锈钢材质，上有香槟色的涂层。帽子架是雕塑的造型，同样是香槟色的不锈钢材质，底座装有荧光灯带。

LOCATION
Tokyo, Japan
DESIGNERS
Architect Ilaria Marelli
COMPLETION
2007
PHOTOGRAPHERS
Nacasa & Partners Inc.

项目地点
日本,东京
设计师
依拉利亚·马瑞利设计事务所
完工时间
2007
摄影师
那卡沙合作公司

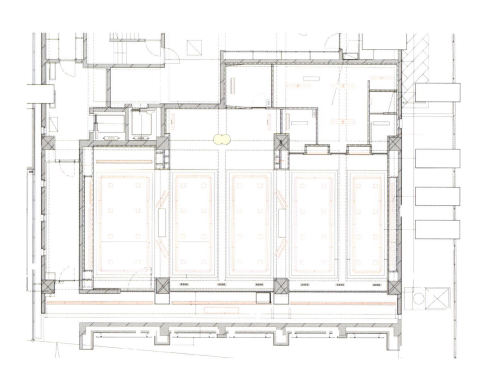

Apartment – The Wonder Castle

魔幻城堡

"Apartment", a luxurious "wonder castle" able to evoke ancient stories and antique objects, is the new design concept for Pitti Bimbo by Ilaria Marelli, dedicated to fourteen selected brands of contemporary kids' fashion such as Mark Jacobs, Bonnie Young, Jean Michel Broc and Fasciani.

Passing the spectacular main entrance with golden oversize lamps and animated by figurants in historic costume, visitors enter the great hall with halberds and shields on the walls, all reminding of ancient and noble ages. From this hall, visitors enter the central fireplace, with room for taking a seat and relax and enjoy the atmosphere, created by the projection of the fireplace and the decoration with shields and deer. The armours in the small mirror-room next door seem like an entire army thanks to the infinite reflections. From here, visitors reach the dining room where traditional dishes are served on the desk of a golden capitonné bar. Next to this scenography of magnificent corridors there are eleven exhibition zones interpreted in different ways as surreal castle rooms. Wooden chairs with long legs are turned into coat hangers in the dining room, white painted old books furnish the "study room", an ancient bathtub and a dressing table adorn the "vanity room", a gramophone and some music symbolise the "music room", and so on also for the rooms of dreams, conversation, wardrobe, kitchens, laundry and small garden. On the walls some of these characterising elements, repeated in numerous pieces, become unusual displays for the kids' fashion.

"奢华的魔幻城堡能够让人联想到遥远的传说和古典的物品"，Pitti Bimbo童装店（意大利知名童装品牌，旗下包括马克·雅各布等品牌）的设计正是秉承这一理念，为顾客呈现一个美妙的购物空间。

主入口在巨型金色吊灯及穿着传统戏服的塑像的装饰下，更加壮观，突显活力。客人从这里进入大厅，墙壁上布满了戟和盾，仿佛走进了古老的贵族世界。从这里出发，便走入壁炉间，享受安详与宁静。旁边的镜室（四壁及天花板都是镜子）内装饰着盔甲，在镜子的反射下，如同一支整装待发的军队。餐厅内，金色的餐桌上摆放着传统的菜肴，长腿的木质座椅可用做衣架。走廊处设置着11个展区，每一个都带有不同的特色，如同城堡内的房间。白色封面的古典书籍为书房增添了一丝特色；古老的浴缸及梳妆台使卫生间个性十足；老式的留声机及乐曲令音乐室别具一格。此外，这里还有闲谈区、更衣室、厨房、洗衣房及小花园等。墙壁上，样式各异的特色元素重复叠加，成为了童装展与众不同的展品。

LOCATION
Florence, Italy
DESIGNERS
Sabine Schweigert
COMPLETION
2009
PHOTOGRAPHERS
Sabine Schweigert

项目地点
意大利.佛罗伦萨
设计师
塞宾·谢尔格特
完工时间
2009
摄影师
塞宾·谢尔格特

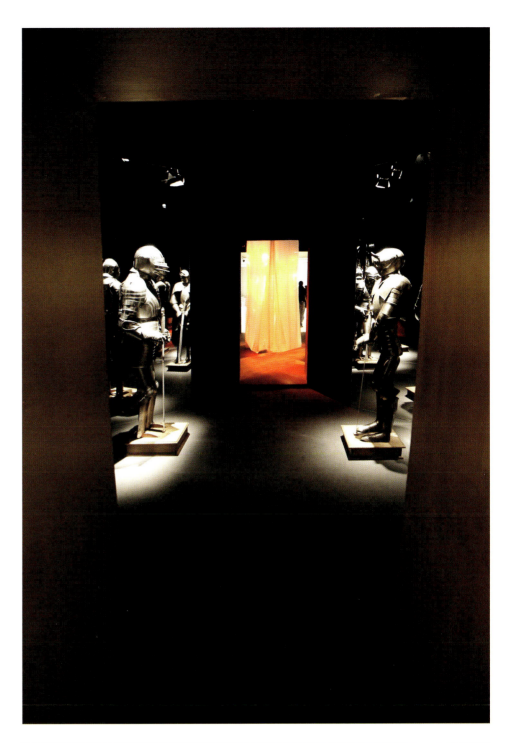

inge van den broeck

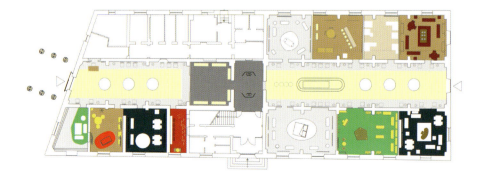

Casa Palacio

卡萨·帕拉西奥

For DIN interiorismo, commercial design is to understand the shopping experience. First of all it is important to know what is to be sold and who are the buyers. Once having perceived the main goal, all the ambiances that will integrate the general concept, are developed to transmit the individual essence of the project. The result is the perfect combination between the day to day activities of the business and the encountering of the buyers that ends in the commercial transaction.

This project was a challenge of coordination, consultancy and stage design for the first store specialised in home furniture and accessories of the Grupo El Palacio de Hierro, one of the most important retail stores in Mexico. The main target was to develop an interior design with the commitment of offering a space where the most important aspect is to feel the atmosphere where lifestyle is the essence.

While browsing through Casa Palacio, the customers discover that each space has a particular lighting, aroma and music, all selected specifically for each area. The transition from one to the other is very soft, as all the areas may be reached from several directions, opening a wide variety of views and different routes to discover the whole project.

The store was designed with experiences more than products in mind, in the interior of a round shape building of about 6,000 square metres. The two levels are connected with electric staircases under the natural light of a magnificent dome.

DIN在商业空间的设计上，一直秉承着"深入了解购物过程"的理念。在确定了商店的类型及潜在顾客之后，制定整体设计概念，并赋予空间不同的特色。

作为帕拉西奥百货公司的第一家家具及食品店，卡萨·帕拉西奥商店在室内空间设计上注重营造一种家居生活氛围，当然这也是设计师面临的最大挑战！

店内的每一个空间都别具特色，拥有各自风格的灯饰、弥漫着不同的气息、播放着不同的音乐。空间之间的过渡缓和而流畅，为顾客呈现出不同的景象。

此外，店内设计更加强调"购物过程"而非商品本身。两层空间通过电梯连接，自然光线从圆屋顶上泼洒下来，明亮而舒适。

LOCATION
Mexico City, Mexico
DESIGNERS
DIN interiorismo, Aurelio Vázquez Durán
COMPLETION
2007
PHOTOGRAPHERS
Courtesy of Casa Palacio

项目地点
墨西哥.墨西哥城
设计师
DIN室内设计公司
完工时间
2007
摄影师
Courtesy of Casa Palacio

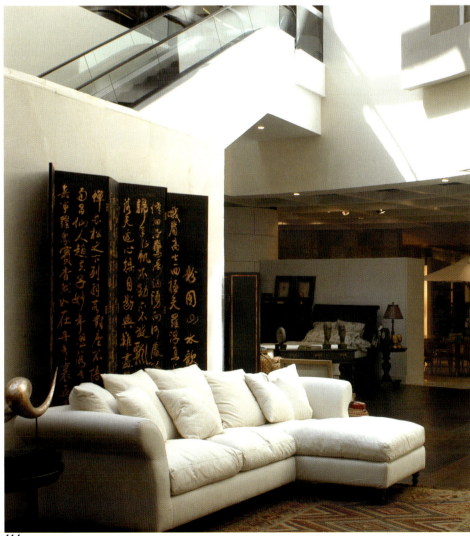

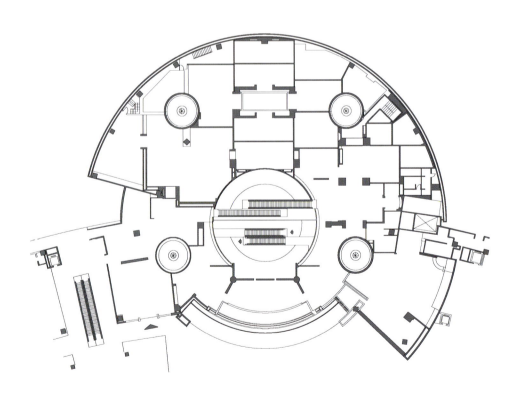

Wynn Macau

永利澳门酒店

LOCATION
Macau, China
DESIGNERS
Wynn Design & Development
COMPLETION
2006
PHOTOGRAPHERS
Wynn Design & Development

项目地点
中国,澳门
设计师
Wynn Design & Development
完工时间
2006
摄影师
Wynn Design & Development

Wynn Macau has 600 rooms and suites. The retail esplanade at Wynn Macau houses exclusive and luxury shopping boutiques in approximately 4,274 square metres of exquisitely designed retail space.

The fountain's performance results in lofty plumes of water and fire dancing to music and permeating the lush surroundings.The centrepiece of the entire atrium is an iconic Golden Tree of Prosperity that rises up from below and dazzles with its golden leaves and branches. The 21 metre diameter gold cupola in the ceiling features twelve sculpted, dynamic animals from the Chinese zodiac. Dramatic, swirling, and transforming patterns are featured in the giant 20 metre diameter LED video screen once the Chinese zodiac cupola opens like an iris, expressing a complexity of rhythms and emotions. The 11 metre diameter sparkling chandelier composes of 21,000 crystals. Echoing the Chinese zodiac cupola is a dome of the western zodiac on the floor. The 10 metre diameter copper dome, engraved with 18th century astronomic charts, features twelve astrological signs. Symbolising vitality, good fortune and well-being, the Dragon of Fortune combines traditional sculptural art, modern lighting and audio effects, in a dramatic display to guests as they enter Wynn Macau's Rotunda atrium.

The meeting and conference areas offer state-of-the-art facilities to ensure that the convention spaces at Wynn Macau exceed guest expectations. The Convention Area is conveniently located on the first floor, accessible by escalator or dedicated elevators. It features 2,200square metres of flexible, multi-purpose meeting space.

永利澳门酒店共有600间客房和套房。占地4274平方米的永利名店街,气派显赫,拥有环球最顶尖、最豪华的购物名店。

动感音乐喷泉那高耸的水柱和火焰在布满翠绿的花园内随音乐节奏起舞。黄金吉祥树破土而出,一柱擎天;金枝玉叶,闪闪生辉,为中庭的焦点所在。金碧辉煌的中庭圆顶直径为21米,铸造了12只栩栩如生的中国十二生肖动物雕塑。巨型LED影像屏幕直径为20米,随着中国十二生肖圆顶如鸢尾花般缓缓盛开,展示一系列变化万千的旋转图像,缔造不同的节奏及情绪。水晶吊灯直径为11米,展现超过21,000颗耀眼水晶。中庭地上的铜圆顶直径为10米,刻上18世纪天文学中的十二星座,与中国十二生肖圆顶相呼应。富贵龙象征龙精虎猛、富贵康乐,是一个集传统雕塑艺术、现代灯光及音响效果于一身的地标,竖立于酒店第二期入口中庭。

永利澳门的会议区备有最先进的设施,为宾客提供专业的会议服务。会议区设于一楼,可经扶手电梯或专用升降机进出,方便快捷。拥有多功能用途的会议场地占地2,200平方米,可供宾客灵活使用。

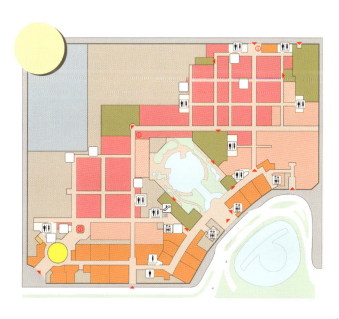

Casal Grande Hotel

卡萨尔·格兰德大酒店

The building area is surrounded by a lovely rural area with small country houses, groups of trees and big vineyards. This rural landscape creates strong suggestions that influenced the whole project.

The idea of the project combines sustainable environmental strategies in a strong architectural language achieved through the use of very contemporary materials, the reinforced coloured concrete and the cor-ten steel. Material is chosen in the light of the harmony existing between them, for their ability to overlap with the chromatic structures alternating in the countryside and because it needs little maintenance and speed of execution-assembly.

The building has been divided into two units which are held together by a passage on the ground floor and very different as regards external surfaces and relationships between blank and solid spaces. The central court has been conceived as an introvert area, and both rooms and public spaces overlook this pleasant background.

The two main architectural elements, held together by a passage on the ground floor, are very different as regards external surfaces and relationships between blank and solid spaces.

The one-floor building contains all the public spaces, namely the reception near the entrance, the big hall, the bar area and the coffee area overlooking the countryside, a cosy meeting area and the service spaces.The main building develops on 3 floors composed by 42 rooms offering a beautiful view on the countryside or on the green court.

Special attention was paid on studying the decorative features, which are characterised by basic lines and made of wengè wood.

酒店的四周环绕着精致的山间小屋、郁郁葱葱的树林和宽阔的葡萄园，浓郁的田园特色赋予其独特的魅力。设计师通过运用现代化建筑材料——彩色混凝土和耐候钢（一种耐大气腐蚀的高强度钢材）打造强烈的建筑语言，同时实现了其倡导的环保理念。

整个建筑在结构上分成两人部分，之间通过一楼的走廊连通。这两部分在外观等方面各具特色。中心庭院则打造了一个低调的背景，从客房和公共区都可以瞥见这里的景象。

一层的结构用于设置所有的公共空间，包括入口处的接待台、大厅、酒吧、咖啡厅、会议区以及其他服务区。主建筑部分共为三层，42个客房，在这里可以尽情欣赏乡村的美丽景色。

此外，设计师在装饰上花费了大量心思，运用简洁的线条和鸡翅木材料创造了非凡的效果。

LOCATION
Reggio Emilia, Italy
DESIGNERS
STUDIO M2R atelier d'architettura
COMPLETION
2008
PHOTOGRAPHERS
Diego Parolini

项目地点
意大利.瑞吉欧·艾米利亚
设计师
M2R建筑工作室
完工时间
2008
摄影师
蒂戈·帕罗利尼

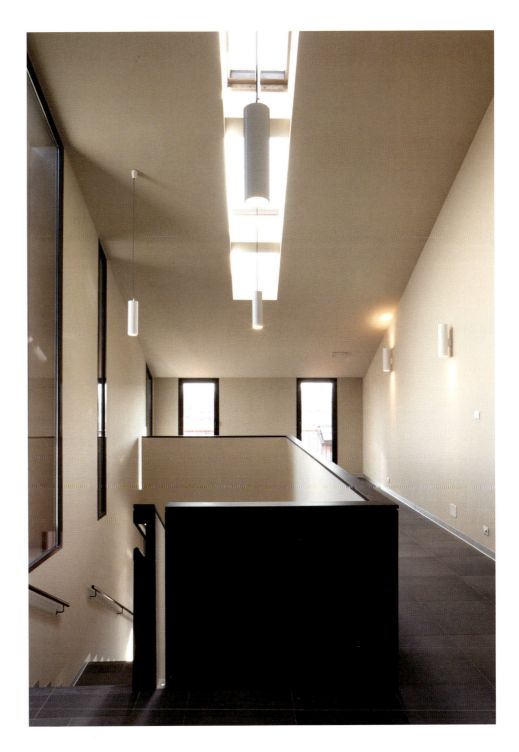

Lánchíd 19

Lánchíd 19设计酒店

Named after Budapest's famed "Chain Bridge" spanning the Danube and situated near both it and the Buda Royal Castle, the Lánchíd 19 has become a contemporary architectural landmark that attracts a new kind of cosmopolitan crowd. Lánchíd 19 is only 10 kilometres from airport and 10-minute walk distance to the city centre.

A remarkable moving glass façade glows with subtly changing light effects; the lobby itself is equally dramatic with its ultra modern Bloomy chairs that seem to float on a glass floor revealing ancient Medieval ruins below.

Principles of flowing space and light dominate the experience of this stylish urban get-away, from the glass atrium towering above the foyer to the rooms' and suites' open-plan design. The feeling of openness and transparency is enhanced by the clear views from the lounge to the restaurant to the intimate garden beyond. The team of Hungarian architects also took every opportunity to exploit the hotel's potential for fantastic views over the city, which can be enjoyed from the suites' terraces or even from the deep bathtub.

The Lánchíd 19 is a beacon of innovation while still paying homage to its historical settings – a perfect point of departure for discovering Budapest's wonders.

Lánchíd 19设计酒店毗邻横跨多瑙河的著名"锁链桥"（并因此而命名）与皇家城堡，现已成为这一地区的当代建筑里程碑，吸引着来自世界各地的人群。酒店距离机场仅有10公里，与城市中心也只有10分钟的路程。

玻璃表面营造出光影变幻的效果，格外引人注目；大厅内超现代的风格的座椅似乎"浮动"在玻璃地面上，特色十足。

流动的空间感和动态的光影效果构成这一风格时尚的城市度假酒店的两大特色，玻璃中庭、大厅、普通客房、套房全部采用开放式格局设计。在大厅内可以清晰地看到餐厅和小花园，为客人营造了一个开阔、透明的环境。此外，客人在套房的露台以及高大的浴缸内都可以欣赏到布达佩斯城的美景，这也是建筑师设计过程中别出心裁的创意。

酒店在注重营造现代风格的同时也充分地利用了本地的历史特色，可堪称是"寻找布达佩斯特色"的旅程出发点。

LOCATION
Budapest, Hungary
DESIGNERS
Péter Sugár, Lázló Benczúr/Dóra Fónagy (D24)
COMPLETION
2008
PHOTOGRAPHERS
Péte Sugur

项目地点
美国,波特兰市
设计师
Péter Sugár, Lázló Benczúr/Dóra Fónagy (D24)
完工时间
2008
摄影师
皮蒂·苏格

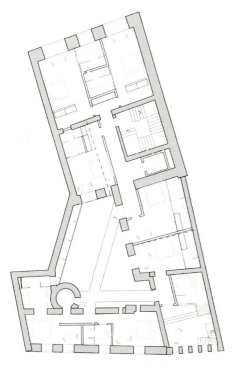

Balance Holiday Hotel

巴兰斯假日酒店

In the heart of the famous "Europa Sport Region Zell am SeeKaprun" affording views of snowcapped Alpine peaks just a few steps from the lake and the skiing area, 10 minutes from local train station and 50 minutes from Salzburg airport, Balance Holiday Hotel is the destination for wellness and recuperation. Its inviting understated modern design not only generates an atmosphere of relaxing elegance, but is sensitively implemented to promote inner balance.

The holistic concept behind the Hotel aims to create a sense of "flowing privacy", a homogenous space that always places the individual at the centre. The 47 unusually large guestrooms are in soothing natural wood and stone. The coulour of each room is different. Every time coming in this hotel, will give you a fresh feel. The atmosphere here is warm and happy. You can relax yourself from the busy daily life. If guests feel particularly invigorated and energised, the surrounding mountains, reaching altitudes of 2,743 metres, eagerly await them.

Hotel has 43 design rooms and 4 luxurious design suites, situated on the top floor of the Design Hotel. The size of the rooms ranges from 32 metres with the Balance Superior room to 81 metres with the Hotel Suite. Guests can choose between rooms and suites with a bathroom separate from the bedroom or with a bathtub integrated into the living area.

巴兰斯酒店位于著名的采而湖欧洲体育中心，与湖区和滑雪场仅有几步之遥，可以欣赏阿尔卑斯山的壮美雪景。这里交通便利，距火车站和萨尔斯堡机场仅有10分钟和50分钟的车程。

设计师秉承"流畅私密性"的理念，打造了统一的空间结构，使顾客成为中心。47间超大号客房全部采用天然石材打造，温暖而亲切。每个房间都采用不同的颜色装饰，给客人新鲜感。暖意融融而又欢快愉悦的氛围让他们从喧嚣繁忙的城市生活中彻底"解脱"出来。如果客人在舒适的环境中备感经历充沛，那么就可以向四周海拔2743米高的山峰"进军"了。

此外，酒店顶层还特别设计有43间普通客房和4间豪华套房，房间宽度从32米到81米不等。客人可以自行选择是入住带有独立浴室的客房还是浴室设在生活区的客房。

LOCATION
Zell am See, Austria
DESIGNERS
Niki Szilagyi, Evi Märklstetter
COMPLETION
2008
PHOTOGRAPHERS
Niki Szilagyi

项目地点
奥地利,萨尔茨堡
设计师
尼奇·斯拉吉 艾维·马克尔斯泰特
完工时间
2008
摄影师
尼奇·斯拉吉

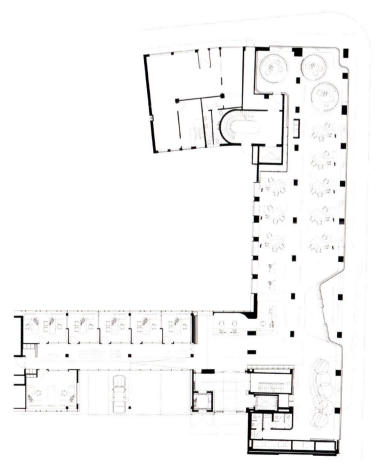

Hotel Rho
Fiera Milano

米兰国际展览中心酒店

A sober and functional building dating back to the early 1900s has been transformed into a contemporary business hotel as the result of an important operation in industrial archeology. Where cotton was once woven, now new business relationships will be woven.

The project brings hospitality to an industrial building, establishing a functional continuity of space. In the former production area of the cotton mill, three floors have been built to accommodate more than one hundred rooms and the tower where the offices and common areas were once located has now become the hotel's public area. The centre of the project, the only element connecting the floors along which the multiple functions of the hotel are spread out, is also the central tower where the public areas are concentrated.

The top floor boasts a convention room, fitness and wellness area; the first floor has a reading room decorated with white lacquered boiserie, while the lobby features an eye-catching red, studded leather reception desk. The common denominator of the space is a dramatic lamp that spans the entire structure and the completely renovated wrought iron floral balustrade that runs the length of all the floors.

Many of the hotel's elements echo the building's original purpose, in terms of function, as already mentioned, style, atmosphere and the choice of materials for the structures and furnishings. The floral wall is designed with a retro sensibility, like the hinged mirror at the head of the bed, interpreting Art Deco and Bauhaus styles with a contemporary flair.

米兰国际展览中心酒店由始建于20世纪初期的一幢素雅而功能性极强的工业建筑改建而来。设计师秉承着"有工业生产的地方必定存在商业发展"的理念以及根据工业考古学的重要研究，打造了这一现代韵味十足的商业酒店。

设计师将宜人的氛围"移植"到原有的工业建筑中，打造了一系列功能连续的空间。纺织厂的生产区被改造成三层、100多个房间的住宿区，原有办公行政区的塔楼（连接各个楼层）现已成为酒店的公共区。

顶层包括会议室、健身房等；一层设有装饰着白色细木护壁板的阅览室；大厅内红色的皮质接待台格外引人注目。此外，酒店内最具特色的莫过于横跨整个空间的大灯以及翻修的熟铁花纹栏杆。

酒店内的众多元素无论是在功能、风格、氛围以及材料选择上都体现出最初的设计意图：花纹图案墙给人复古的感觉，同床头的镜子一起打造出散发着现代气息的艺术混搭（ART DECO）和包豪斯学派风格。

LOCATION
Milan, Italy
DESIGNERS
Caberlon Caroppi Hotel&Design
COMPLETION
2008
PHOTOGRAPHERS
Lorenzo Nencioni

项目地点
意大利.米兰
设计师
砍博伦+卡欧皮酒店设计工作室
完工时间
2008
摄影师
洛伦佐·嫩乔尼

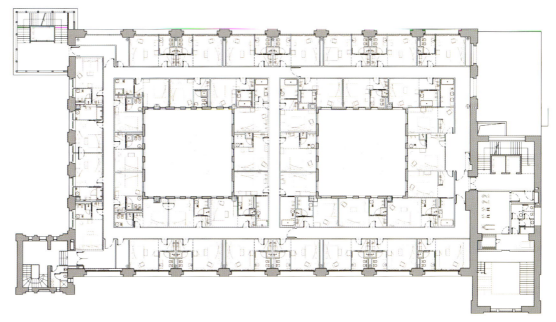

Barry

巴里小学

The new Commodore John Barry Elementary School accommodates 646 students from pre-kindergarten to the 8th Grade. The school is built on site of the old school, which is located at 5900 Race Street. The site is 4,173 square metres and is bounded on all four sides by two-story brick row houses, dating to the early 1900s.

The design challenge for this project is to provide a first rate academic facility on a tight urban site. To maximise outdoor play area and neighbourhood green space, a four-storey school was designed.

The ground floor is 2,383 square metres, preserving over 40% of the site as open space. This space is being developed for early childhood play areas, as well as hard and soft surfaced play areas for older children.

The design divides the school into three vertical zones. The base zone houses functional spaces including the lobby, cafetorium and administrative offices. These create a public commons for the students and a meeting place for the community.

The middle zone consists of two identical floors that contain grade assigned classrooms. The classrooms are arranged to create smaller internal communities, the "small schools within a school". The two-storey Gymnasium occupies the centre of the middle zone.

The top of the school is developed as a special learning centre. Classrooms for arts and science and the instructional media centre share outdoor decks that can be used for hands-on learning experiences. A special education classroom and outdoor play area is also located on this level.

这间新建的约翰·巴里海军小学可容纳从幼儿园至8年级的学生646人。学校建于其旧址——士街5900基础上。建筑地点占地4173平方米，四面濒临一栋建于20世纪初的双层砖瓦排房。

项目宗旨是：在郊外建立一所有着一流教学设施的学校。最大限度的扩大学生的游乐场地和绿地面积。于是，这所四层楼的学校应运而生。

建筑底层约2383平方米，且有至少40%面积为开放空间。这里用来作为幼儿及稍大一点孩子的游乐场所。

设计师的设计将学校分为三个垂直空间。底层空间包括：大厅、兼作礼堂和自助食堂的大厅、行政管理办公室。这里为学生提供了公共集会场所及社团聚会地点。

中部的空间包含一系列不同年级的教室。占有两层楼的面积。这些教室的设计创造出小型室内效果，也可称为"学校中的小学校"。其双层的健身房占据中部空间的中心位置。

学校的顶层空间是一个特殊的学习中心。那些用作美术、科学研究以及媒体教育中心的教室，其课桌都被安置在户外，用以作为手工学习的场所。一间特殊教育教室和一处户外游乐场所也坐落在本层。

LOCATION
Philadelphia, Pennsylvania, USA
DESIGNERS
Ross Barney Architects
COMPLETION
2006
PHOTOGRAPHERS
Matt Wargo

项目地点
美国,宾夕法尼亚州 费城
设计师
罗斯·巴尼建筑事务所
完工时间
2006
摄影师
马特·瓦戈

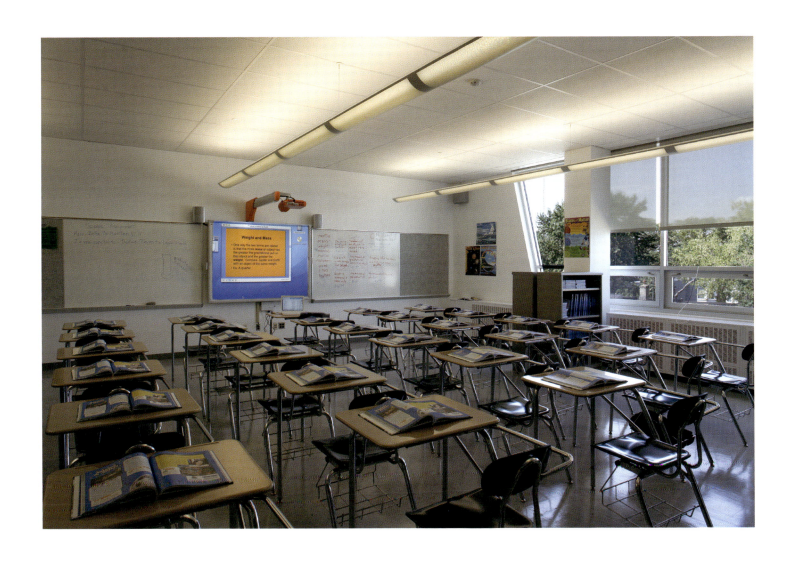

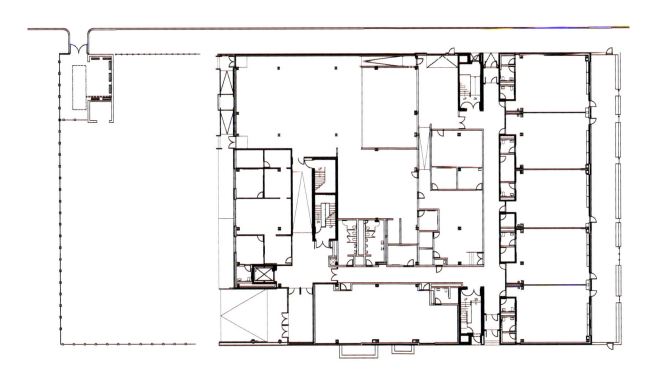

Mark Taper Forum

Mark Taper Forum话剧院

One aim of the project was to achieve both functionality and preservation objectives. Rios Clementi Hale Studios and theatre consultant Sachs Morgan Studios worked together to respectfully update the building. Outdated and inefficient spaces and systems were identified, evaluated, restored, or renovated to be compatible with historically significant elements of the building. The updated interiors are directly influenced by the original glamour of the New Formalist design and were inspired by a distinct and precise tradition of craftsmanship and detailing of the original building. New finishes are modern and classic, and pay homage to the beautiful form and geometry of the 1967 Welton Becket-designed building. The renovation brings the building into an era where big productions are the norm and sensitivity to patron needs is required to remain relevant.

The renovation allows for additional and more-spacious interior areas for the Taper's audience and artists. By reducing the ticket booth size, raising the lobby floor to be flush with the exterior ground plane, and moving the restrooms downstairs, the architects open up the lobby to present a more fitting entrance showcasing the original Tony Duquette abalone tile wall.

The new lounge is a comfortable, contemporary space. Rios Clementi Hale Studios gained additional square footage by utilising a portion of the underground parking garage to extend the theatre space, adding plentiful room for the 125-square-metre lounge. The broad curves of the walls reflect the overall shape of the building, giving the illusion that the lounge existed originally.

设计的目标之一即要在保留原有建筑风格的同时，满足功能性的要求。里奥斯·克莱门特·哈勒建筑工作室同剧院顾问共同合作，将老旧的空间及低效的系统或被保留，或被翻新，延续了历史特色。设计师在新形式主义风格及手工制作传统的双重影响下，打造了一个古典风韵十足而又不缺乏现代感的室内空间。

设计师减缩了售票亭的规模，把休息室移至楼下，之后将大厅升高到与室外地面相平的高度并将其打通，将原始的陶瓷墙突显出来。

新建的休息室风格现代，舒适温馨。设计师利用了地下停车场的部分空间，扩大了原有的休息室的规格。墙壁上宽阔的曲线造型展示着整个建筑的形状，给人一种错觉——休息室似乎一直存在。

LOCATION
Los Angeles, USA
DESIGNERS
Rios Clementi Hale Studios
COMPLETION
2008
PHOTOGRAPHERS
Tom Bonner, Craig Schwartz

项目地点
美国.洛杉矶
设计师
里奥斯·克莱门特·哈勒建筑工作室
完工时间
2008
摄影师
汤姆·邦纳.克雷格·施瓦兹

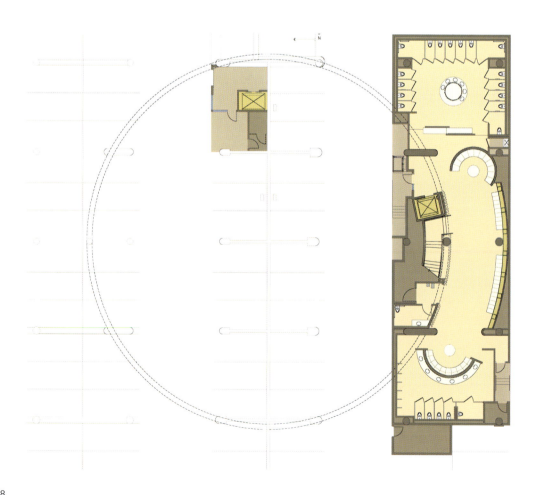

Children's Hopsital at Montefiore

蒙特弗洛尔儿童医院

When the hospital received a grant to build a new chemotherapy infusion lab for children, they asked Rockwell Group to rethink the way to deliver oncology care. The design began by choosing a material and colour palette that would be fresh and optimistic, but sufficiently sophisticated to address the patient's wide age range. Rockwell Group believes strongly in finishes that are narrative, not merely decorative. Because existing theme of the floor was "we are all star stuff", the designers created a series of custom floor tiles that celebrate biodiversity with the silhouettes of plants, minerals, animals and marine life. A central lighting element features the dancing forms of "birds of the Bronx", from the winged dinosaurs of 50 million years ago to the exotic species that live nearby in the Bronx Zoo.

The biggest innovation is the custom bay divider. Because space did not permit building in a dedicated playroom, the designers sought to bring elements of play into the bays themselves. While the base of the divider provides storage for patients' personal items, the upper portion is a series of translucent fins that can be closed for privacy or opened to encourage social interaction, because kids on a regular chemotherapy schedule often make friends in treatment.

The middle zone of the divider is an interactive dashboard including a wireless laptop, magna-doodle write-on, and rubber-band magazine rack. The front of the bay incorporates games for kids of all ages, arranged by height.

蒙特弗洛尔儿童医院拟建立一个新型儿童门诊输液室，罗克韦尔集团坚持设计富有内涵，而不仅仅是装饰外表。设计选料着色力求体现新鲜的视觉效果和彰显乐观的精神，以充分满足各年龄段患者的审美。借"我们都是明星"这样一个主题，设计师选取了有着动植物、矿物以及海洋生物轮廓的系列锦砖，使得空间极具生命力。拥有"布朗克斯之翼"的灯饰是空间照明设计的一大特色，这一设计将距今5000万年前的翼龙以及布朗克斯动物园外来物种的羽翼刻画的淋漓尽致。

设计最大的创新点表现在输液隔离室的设计。由于空间的局限性，输液都安排在各个隔间里。在分隔板底层设计了一个用于存储患者私人物品的小箱；上层空间被设计为一系列半透明的鳍状构架，可随时闭合，保证个人隐私，也可将其开启，方便定期治疗的孩子互动交流、结伴成友。

隔间的中心地带被设计成一个互动平台，其中有一台无线上网笔记本电脑，一面大型涂鸦板可随意写画涂擦，还有橡皮圈期刊架。此外，隔间的最前端为各年龄段的孩子开辟出一块活动基地。

LOCATION
New York, USA
DESIGNERS
Rockwell Group
COMPLETION
2007
PHOTOGRAPHERS
Rockwell Group

项目地点
美国,纽约
设计师
罗克韦尔集团
完工时间
2007
摄影师
罗克韦尔集团

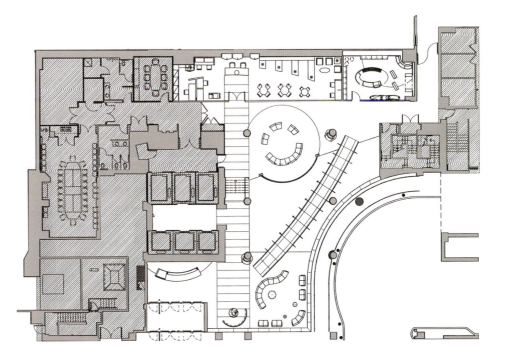

Lyric Theatre

抒情剧院

Lyric Theatre is a professional summer stock company founded in 1963 and the only professional musical theatre in Oklahoma.

The building had been renovated so many times since 1935 that there was no historic character remaining. Historic preservation was not an option. The Architect chose instead to acknowledge the "ghost" of the past. A neon outline "ghost" marquee was designed in the spirit of the original. Exposed brick walls acknowledge the building shell construction and make art of heater cavities and conduit locations once hidden behind long ago destroyed plaster. A new wood lath ceiling adds to the raw character of the entry lobby. An original terrazzo ramp remains and connects outside to the inside at the entry.

The theatre outer lobby was designed to have changing light color that corresponds with the current performance. Plasma screens provide changing donor recognition. The outer lobby is punctuated by 162-centimetre-tall changing LED lighted letters. Toilets remain raw with exposed brick and structure and use coloured light to continue the theatrical qualities. The theatre space maintains the original riveted steel bow-string trusses and includes the addition of lighting balcony, catwalks and a tension wire grid above the stage. Exposed brick marks the original building shell.

抒情剧院成立于1963年，是俄克拉荷马州唯一的专业音乐剧院。

该建筑自从1935年以来被翻新过多次，已经没有历史遗留痕迹的特点了。建筑设计不是对它进行历史保护，而是选择对过去的"幽灵"的理解。基于这个想法设计出一个用霓虹灯勾勒出来的"幽灵"帐篷。用裸露在外的砖墙做建筑外壳，使多年前藏在破碎的石膏后面的暖气和管道的位置成为一种艺术。一个新的木条天花板增加了大厅入口的天然质感。原来就有的一条磨石子地斜道仍然保留，把外面和入口处连在一起。

按照设计，剧院的外部人厅要随着演出的剧目改变色彩的光度。提供鲜红的屏幕来改变捐赠者的认可。外部大厅被几个不断变换颜色的8寸高的LED大字不断打断。厕所的外墙裸露在外，保持天然的风格，同时也用彩灯来继续剧院的特点。剧院空间保留原有的用铆钉钉牢的弓弦构架，增加采光阳台、狭窄过道和舞台上方的有张力的金属网格。裸露在外的砖块标志出原来的建筑外壳。

LOCATION
Oklahoma City, Oklahoma, USA
DESIGNERS
Elliott + Associates Architects
COMPLETION
2007
PHOTOGRAPHERS
Elliott + Associates Architects

项目地点
美国.俄克拉荷马州俄克拉荷马城
设计师
埃利奥特联合建筑事务所
完工时间
2007
摄影师
埃利奥特联合建筑事务所

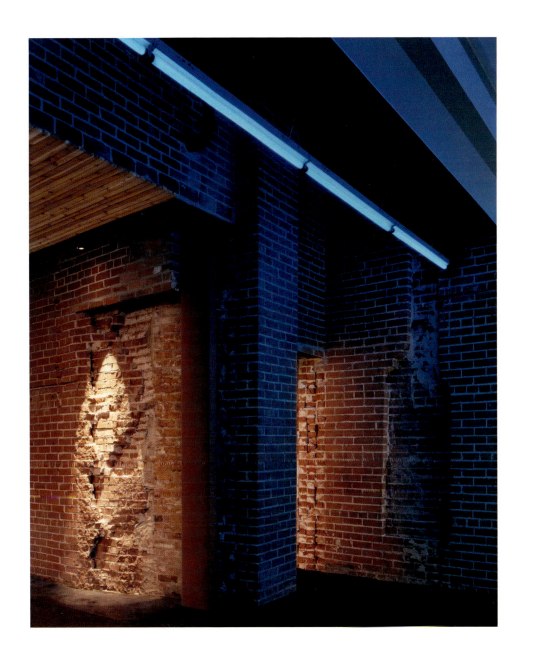

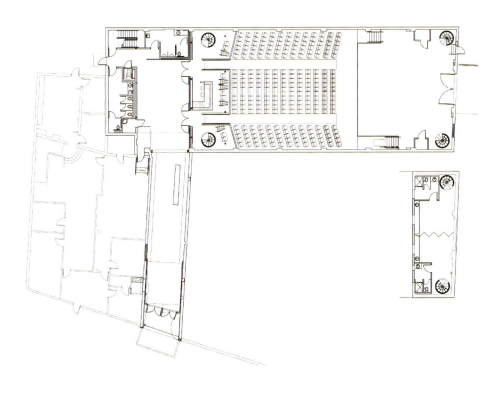

New Office Design–Application of Curves and Other Elements

From Morningstar

新办公空间——曲线及其他
从晨星说开去

文：谢昕宜

Nowadays, the office composed of regular rooms and furnished with rectangular desks is no longer popular. Instead, the space which highlights freedom, liveliness and diversification has been hotly pursued and gradually becomes prevalent in a period that the "human-centred" concept is highly promoted. Based on this, the designers fully explore their imagination and manage to create more and more distinctive office spaces!

"Quiet but not suppressed, relaxed but without noise", that is the basic requirement of an office design. The "Morningstar", a unique corporate headquarters servicing the Asia Pacific Region, catches our eyes and will be exemplified in detail in the following text.

In Morningstar, openness is conceived as the main concept with only the glass panels and furnishing and floor decoration to divide and differentiate the functional spaces. Moreover, the idea of open plan brings as well several actual benefits, avoiding formation of irregular space, solving the problem of day lighting, promoting direct interaction between employees and so on.

As for the colour scheme, the designers select natural colours rather than bright, intense ones so as to create a cozy, home-like atmosphere. Small meeting rooms and resting rooms provide more opportunity for casual discussions.

Other than the above-mentioned, it is the effective application of curves that makes the office itself perfectly stand out. Fluid and tender, curves often have

四方的办公室，四方的办公桌，中规中矩的摆放方式？不，这样的办公空间已经成为了过去，在这样一个人性化的时代里，办公空间的设计已经越来越趋向于自由、活泼、多元，设计师们用自己的画笔，努力为我们打造出一个又一个富有特色的办公空间。

办公室是一个需要安静，又不应压抑，需要放松，又不应喧嚣的地方。在各具特色的办公室设计中，笔者留意到了晨星的设计。

晨星的空间十分开敞，只用玻璃隔断以及设施摆放和地板装饰来区分不同功能区。空间的开敞，即避免了过分切割造成的异型空间，解决了采光等问题，更直接加强了内部的交流，带来全新的工作氛围。

因为空间较大，晨星的设计并没有用到太过浓烈的颜色，一系列中性色营造出家居一般自然舒适的氛围，舒适的小型会议室，休息处，给了员工更多讨论交流的机会，这也是新的办公方式所造就的新办公空间。

然而，晨星区别于其他设计的最大亮点，是设计中曲线的完美应用。曲线可以打破沉闷，同时也流畅温和，曲线可以创造多元，同时也简单利落，这不正是一个办公空间所需要的吗！

晨星的设计从平面布局开始，就已经定下了曲线设计的基调。圆形的建筑平面里，所有空间分区都围绕着中心功能核布置，这样的平面布置，让人想到了这样一个词——团结。这也正是一个企业最最需要的精神。弧线、圆的元素从平面开始，延伸到设计的每一个层面中去，更小范围的布局，比如会议室、休息区、桌椅的摆放都自然而然的形成一种围合感，划定了空间范围的同时，让每一个小空间都具有凝聚力。

在晨星的内部设计中，也延续着平面布局中的圆形元素。一道弧线的墙，一个弧形的隔断，让空间的视觉感受顿时丰富起来。曲面之中又有变化，沿着中心区域外的弧形隔断，厚重的墙体用木材装饰，辅以灯光效

the capability to prevent dullness and create diversity.

All the functional spaces flow along the perimeter of a central nucleus structure – the special layout further forms the foundation for the adoption of curves as well as embodying the spirit – solidarity that an enterprise firmly pursues for and adhere to.

Besides the overall layout, curved forms continue to appear in the interior design. Curved walls and panels in no doubt enrich the visual effect of the space; the wooden surface of the wall shinning under the dramatic light creates different views as one moves along; partition structure painted brick-red stretch between the ceiling and the floor, forming a series of private areas; the selection of round tables and lamps corresponds with the overall layout and further enhances the design concept.

If the curves embody the characters of sophistication and flexibility required in doing a job, then the straight lines that connote honestness and frankness are equally necessary. In Morningstar, straight lines compliment the whole design where curves play the main part.

In a time that creativity, communication and casualness are highly prompted, we have already witnessed the efforts the designers spare in constructing new-style offices and we believe they will bring us more and more distinctive designs!

果，沿着墙壁可以感受移步换景的视觉效果。砖红色隔断或斜向联通天花板和地板，或在上部和天花板之间留有空间，流畅的造型圈出了一个小小的私密区，无论从哪一个角度看都是这个空间里出彩的一笔。再细化下去，还有圆形的桌子和灯具。

Engine的设计有异曲同工之妙，环状沙发、圆形茶几、圆形灯具在上方相呼应，搭配和谐而优雅。而在"丽思塔办公"中，斜拉的白色条状装饰也同样构成了曲线的隔断，只是比晨星的设计更多了通透感而已。线条有曲直，做人亦如此，如果说曲线如同职场上的圆润通融一样惹人喜爱，那么直线就如同耿直刚正一般不可或缺。如果说晨星室内的曲线设计得益于平面布局，那么直线设计就是突破整体曲线环境之中的点睛之笔。晨星的总体布局，在层层同心圆中，又有径向直线的加入，弧形是向心，而径向的直线则大有飞扬发散之感。内部物品摆放又较为简约，或方或圆，没有与圆形的平面造成冲突，反而更好的突出了弧线设计的妙处。

在其他的设计中，也可以看到设计师为创造更为舒适的办公空间所作的努力，在这个创造的时代，在这个交流的时代，在这个自由的时代，为我们带来更多特别的办公空间。

Homey Office

办公如家

文：又清

In recent years, SOHO, a brand-new office style, has gradually become prevalent so that sometimes it is not easy for us to differentiate office and home. Office is often flooded with elements commonly used in home design or vice versa. An obvious example is the use of bright colours.

It is with no doubt a rather bold idea to employ colours such as pink, red and purple to decorate office and of course can not be readily accepted by most people. But now it has been extensively realised and hotly promoted. In Engine and Banco Deuno, both of them adopt such colours for decoration, rendering the space replete with warmth and intimacy. In Engine, the entrance lobby forms the focus of the entire office where pink is largely employed and irregular curves create floral patterns on seating area. Such design combines both functional and artistic features as well as bringing in homey feel for visitors upon entering the office. In Banco Deuno, it is the reception space that quite catches eyes – in the background of red and violet, the vividly floral-shaped curves make the small space more warm and welcoming.

As for the above-mentioned, however, we can not pass over an important fact – bright colours are usually used in public space in office design, for in working areas the designers still prefer calm colours to create a tranquil atmosphere.

Moreover, not the same as colours, the homey feel is widely pursed in the design of the entire office. LG European Design Studio is a solid example. In a space where openness and rigidity dominate the whole office, the wooden floor changes the main theme and brings in a touch of intimate feeling, further reflecting the LG spirit – we are family members. Additionally, the washes of light green feature the whole design. In an office where natural colours flourish, the light green instantly attracts visitors' attention and certainly makes the space more bright and lively – the seating chairs painted green are decorated with straight lines and scattered everywhere. Anther two examples are Morningstar and Google Office. In the resting room of Morningstar, the sofas of irregular shapes are brought in – imagine seating there and immersed in the soft lights, how pleasant it is enjoying a cup of tea or just thinking!

In Google Office, sofas in sandbag shape and a traditional ship compliment each other in the conference room. Besides, cable cars are transformed into working areas and resting rooms of various styles, forming a layout of "room in room". Decorated with cotton, lace and leather, those small spaces can inspire more creative imaginations.

In the end, we must recognise that design in nowadays has no boundaries. With their luxuriant imaginations and exquisite senses, the designers can bring us more and more distinctive spaces!

当"新办公""家庭办公"等词汇逐渐变成热门，有时候看到设计，我们会不自觉的想，这是家居设计，还是办公？SOHO一族的家里会带着办公室明亮开敞的感觉，而办公室的设计中，也充斥着曾经独属于家居设计的私人元素。比如色彩。谁能想到粉红、玫瑰红、紫色这样的颜色出现在办公空间中？可是Engine和Banco Deuno做到了，这两个办公设计都用到了这样温暖而私人的颜色，而通过设计师巧妙的设计，这样的颜色不但不会让人觉得不合时宜，反而会让人更觉得亲近。

Engine的入口设计是一大亮点，也是漂亮的粉色最集中的所在，坐凳用不规则曲线勾勒出花的形状，流畅而自然，上方造型特别的灯与之呼应。这样的设计既有实用性，同时也别具一格，让人一进门就有如回家一般的归属感。Banco Deuno的设计有异曲同工之妙，不同之处在于Banco Deuno是一个更小更紧凑的空间，这个空间里同样用到了温暖的红色系，玫瑰红和紫罗兰的背景，加上可爱的花朵形状曲线，让小小的接待空间充满亲和力。

当然，这些都只是偏向公共的区域，温暖和活泼的设计可以带给人愉悦的感受，然而办公空间作为一个工作的地方有着他独特的需要，在真正的工作区，设计师的用色则转为沉敛，营造出一个整洁安静的办公区。

也有将家一般的感觉贯穿于整个办公空间的设计的，比如LG欧洲设计中心。打开图片，就忍不住惊呼：竟然用了家装系地板！LG欧洲设计中心的整体设计风格空旷而硬朗，充满电子从业者干脆利落的感觉，但是就是这样一个地板的细节，让这个办公空间增添了亲和感，因为不论公司有多大，都让人觉得我们是一家人。LG欧洲设计中心室内设计的妙处不仅仅于此，更让人喜欢的是设计中那一点鲜亮的绿。当看见平面图上这些地方都用绿色标明，不禁会心一笑，果然是这个设计的特色所在。家装一般的中性色中，这样一点鲜亮的绿，立刻让整个空间明快起来。那些绿色家居的造型也都别具一

格，最吸引人的莫过于那些散落的座椅。硬朗的线条勾勒着座椅的每一个边角，然而那些线条又不是中规中矩的横竖直，而是不规则的倾斜着，让整个座椅的感觉简单而充满趣味性，又或者像一个吸引人的奇妙幻想，为空间增添了活力。原本可以用曲线代替的座椅设计，设计师依然执着的选择了棱角分明的直线，这样的手法在LG欧洲设计中心的设计中随处可见，不知道这样设计的背后，是否也代表着LG员工们在工作与处世中直来直往的特点？晨星的设计中更为大胆的用到了舒适而私人的家具——在晨星的休息室中，摆放着那样不规则形状的沙发，光线柔和的休息室中，坐在这样的沙发上品茶或者冥想，都应该是极为美妙的吧。

这种沙发同样出现在大名鼎鼎的Google办公设计中，沙袋一样的沙发与一只旧船巧妙的结合，创造出一个可爱的小小会议室。Google办公设计的新奇之处远不止于此，带有loft气质的旧缆车改造，被设计成各种各样不同风格的办公或休息室，无论是舒适的棉布，迷人的蕾丝，还是豪华的皮草，与停放在室内的小屋子相结合，都可以创造出惊人的效果，如此自由而富有想象力的空间里，自然可以大大激发Google公司中那些激情澎湃着的员工，为更多的人创造更多愉快的网络生活。

说到这里，也许早已超越了家居设计融入办公设计中的温暖与自然，而是更为自由，更为舒适的空间，其实我们的设计早已没有了界限，只是设计师在用他们丰富的想象力与敏锐的心，在为我们打造一个又一个，距离心灵最近的空间。

Greutmann Bolzern

Greutmann Bolzern工作室

Defining an Office as a Social-physical System

An interview with Greutmann Bolzern

办公：兼具社会性和物质性
——访Greutmann Bolzern工作室

In 1984, Carmen Greutmann-Bolzern and Urs Greutmann founded the Greutmann Bolzern Design Studio in Zurich. Their partnership dates back to university days. The trained structural engineering draughtsmen jointly attended the Zurich University of the Arts (ZHdK). She graduated as an interior designer, he as an industrial designer. Today the couple – together with a team of five colleagues – drafts, plans and conceptualise in all areas of design. The office world is their most important field of activity. Lista and Belux are amongst the clients from the office furniture industry. In addition, the designer couple designs furniture, consumer goods and lamps for companies such as Audience Systems, Dietiker, MHT, Rolf Benz, Röthlisberger, Swiss and Wogg. The studio's works have been repeatedly awarded. Since 2003 Carmen and Urs Greutmann-Bolzern have been sharing a professorship for product design at the Academy of Fine Arts in Munich.

1984年，Carmen和Urs Greutmann在苏黎世建立了Greutmann Bolzern 设计工作室。如今，它已成为瑞士最别具一格的工作室。Carmen和Urs Greutmann是专业的建筑绘图师。这对设计组合的合作可以追溯到大学时代：1980年到1984年，他们共同毕业于苏黎世艺术设计学院。Carmen当时的专业是室内设计，Urs Greutmann则是工业设计。如今，由Carmen和Urs Greutmann带领的这个5人团队涉及造型设计的各个领域。他们的设计因造型和材质的简洁明了而独树一帜。这个设计工作室最主要的设计对象就是办公室，从照明隔板到公司标志。"我们对这个设计主题尤为感兴趣，以此来布置空间，并产生一个通往建筑的交叉点。"像Denz&Co.，Lista和Belux这些办公家具公司也是他们的客户。此外，这对设计组合还为Audience Systems，Dietiker，MHT, Rolf Benz, Röthlisberger, Swiss和Wogg公司设计家具、日用品以及灯具。他们的设计多次获得国际奖项。自2003年以来，Carmen和Urs Greutmann还担任慕尼黑艺术学院产品设计的教授职位。

1."Crossover" is very popular in various design areas nowadays. What is your view towards this and are there such design elements in your works?

Crossover is a trend and like all trends it is limited to time. When defining "crossover" as a term of thinking and acting beyond limitations of any kind, it is a basic condition to be creative at all.

2. In your point of view, what kind of a role materials should play in design?

Today digital media have one great disadvantage: they only show material as a surface, without reflecting the character of haptics, structure and its nature. This results in surface images without any depth – what eventually is also missing in the final product.

3. Details usually determine the overall effect of a design, and what is your attitude towards this?

Good design is the perfect orchestration of details, even though, there are compositions which stand above the detail – good designs and creations which are strong enough to bear defects.

4. A design would involve different aspects.Which aspect do you usually start with? And which aspect is the most important for you?

Our most important aspect is creating reasonable products - which is not necessarily equal to usefulness or functionality.

5.How would you balance the functional and artistic features of an interior space?

In our interior work the artistic aspect becomes more and more important. Interior space needs - next to functionality - further dimensions to maintain validity and identity.

6. How would you deal with the different opinions between your design and a client's demand?

Finally, it all comes down to man. It means that we only achieve good results when we are able to reach them together - as a team.

7. In your opinion, what's the difference between interior design of offices and that of other types? And what are the common features they share?

Nowadays, the essential difference still consists in defining an office as a social-physical system.

8. In your view, how should a human-centred design find its way in an office?

When looking at design in a working environment context, it is definitely the human being who has to be focused

1. 现今，"跨界"开始在不同的设计领域流行，您对此有什么看法？在您的作品中会运用此类设计元素吗？

跨界，可以说是一种趋势，和其他趋势没什么两样，都会受到时间的限制。如果将其定义成为不受任何因素限制的思维和行动，那么就必须学会创新。

2. 在您看来，材料在设计中扮演着什么样的角色？

如今的数字传媒存在一个极大的缺陷——他们只是将材料作为表面装饰元素，并为展现其结构、本质以及触觉等特性。从而导致了"材料只是一种表面元素"的错觉。

3. 细节往往能够决定整体设计，您如何看待？

一个好的设计必定是各种细节的完美融合，即使有一些好的设计和创意足以承担一定的缺陷。

4. 一项设计往往包含不同的方面。您通常从哪里着手，比较注重哪一点？

对于我们来说，最重要的就是创造令人满意的作品——当然不仅仅是指功能性。

5. 在室内空间设计中，您如何平衡实用性与艺术性？

在我们的室内设计作品中，艺术层面的东西变得越来越重要。室内空间，功能性不再是唯一的要求，还需要赋予其特色。

on. Additional, you have to understand that design functions besides only considering the shape. Design allows (or prevents) social activities and interaction. These are things which overcome the design itself. For instance, when creating a park bench you have to not only think about the design but also about what it can do for the public human being and what rituals it may enable.

9. Would you share with us several design details that you find the most satisfying?

A good detail enriches the overall shape of a product without being flashy. It's about what is necessary and what actually makes sense. Nevertheless, it is the handling of details which gives the whole concept is uniqueness.

10. Compared with traditional office designs, contemporary ones focus more attention on non-working areas. Would you talk about your experience and feeling towards the design in these areas?

This development shows that we´re actually longing for more informal, uncomplicated situations. We all have a great yearning for freedom and at the same time we want to be socially integrated. The new interior concepts imply this expression and allow a more open mental interexchange.

6. 如何处理设计与客户需求之间的矛盾？

这就涉及到了人与人之间的沟通。我们会倾尽所能达到令双方最满意的结果。

7. 在您看来，办公空间的室内设计同其他类型空间有何异同？

本质的区别就是办公空间仍被定义成一个兼具社会性和物质性的场所。

8. 您觉得如何在办公空间的设计中实现"以人为本"？

办公空间内，人无疑是主角。此外，必须明白，设计不仅仅只是涉及空间格局，还要考虑到人与人之间的交流。换句话说，很多东西往往超越了设计本身。举个例子，如果想要在公园内设计一个长凳，那么不能只考虑设计本身，还要想到这能够给公众带来什么。

9. 能够谈谈您最满意的几个设计细节吗？

一个好的细节能够让整体更加完美，对细节的处理会让整个理念更加独具特色。

10. 与传统的办公设计不同，现代办公空间更注重非工作区的设计。能对此谈谈您的经验和看法吗？

这一趋势恰恰表明了我们对那种非正式、简约环境的需求——我们极度渴望自由，同时更渴望与人交流。